OTHER SCIENTIFIC AMERICAN FOCUS BOOKS
The Structure of the Universe
Medication of the Mind
Cosmic Collisions
The Language of Animals
Of Mind and Body

THE NEW AGE OF COMMUNICATIONS

JOHN O. GREEN

Foreword by Paul Gilster

A SCIENTIFIC AMERICAN FOCUS BOOK

An Owl Book

Henry Holt and Company
New York

HOUSTON PUBLIC LIBRARY

Henry Holt and Company, Inc.
Publishers since 1866
115 West 18th Street
New York, New York 10011

Henry Holt® is a registered trademark
of Henry Holt and Company, Inc.

Copyright © 1997 by John O. Green
All rights reserved.
Published in Canada by Fitzhenry & Whiteside Ltd.
195 Allstate Parkway, Markham, Ontario L3R 4T8.

Library of Congress Cataloging-in-Publication Data

Green, John, 1945–
The new age of communications / John Green.—1st ed.
   p.   cm.—(A Scientific American Focus book)
Includes bibliographical references and index.
1. Communication— Social aspects.   2. Mass media—Social aspects.
I. Title.   II. Series.
HM258.G695   1996
302.2—dc21   96–46404
                                                                              CIP

ISBN 0-8050-4026-9
ISBN 0-8050-4027-7 (An Owl Book: pbk.)

Henry Holt books are available for special promotions
and premiums. For details contact: Director, Special markets.

First edition—1997

Photo credits appear on page 144.

Editorial and production services: G&H SOHO, Inc., Hoboken, NJ

Printed in the United States of America
All first editions are printed
on acid-free paper. ∞

10 9 8 7 6 5 4 3 2 1
10 9 8 7 6 5 4 3 2 1 (pbk.)

CONTENTS

6
Foreword

8
Introduction

18
Chapter One
Scenarios with a Metabook
Anyone, Anything, Anytime,
Anywhere

27
Chapter Two
Something Is Bound
to Come of It

49
Chapter Three
How the World Is Wired

65
Chapter Four
The Net, the Web,
and the Highway

75
Chapter Five
What's New About
New Media?

102
Chapter Six
The AI Wildcard

120
Chapter Seven
The Vector of Desire:
The Ultimate Telephone

134
Timeline

140
Further Reading

141
Index

144
Photo Credits

F O R E W O R D

hether we call it a digital revolution, an information superhighway, or a grand convergence of media, something is happening to our world that is protean and profound. It was in the 1960s that Marshall McLuhan taught us to think about media not so much in terms of what they delivered—content—but in terms of how their presence changed the way we live. It took the Internet to clarify McLuhan's insight, but every time we send electronic mail or log onto a World Wide Web site, we're witness to the transformation in daily living that computers have created. And we'll never be the same.

But digitization runs deeper than today's World Wide Web. We've learned that a voice, an image, a book, or a song can be turned into the binary 1s and 0s of digital information as readily as the data on an old-style reel-to-reel tape. Suddenly the communications network, which encompasses our fiber-optic long-distance connections, undersea cables, satellite downlinks, and microwave relays, becomes a vast facilitator. We click with a mouse to link to information in other countries. We pull music and images from disc-based encyclopedias. We journey through 3-D digital environments to play games and analyze data.

A sea change in daily life invariably inspires hype, but the great switch foretold by Nicholas Negroponte, so ably explained in this book, is a genuine paradigm shifter. Wire-based telephone traffic takes to the air through ever increasing cellular options, while formerly broadcast media like television burrow into coaxial cable connections. Corporate mergers accelerate as companies fight to adapt to a demand-based model. Internet telephony challenges conventional dial-up calling, while cable television companies try to become cyberspace providers. Movie studios link with data carriers to build tomorrow's hybrid media.

Putting these trends in order is not a task for the timid, for it requires an understanding of technologies as diverse as telephones, mainframe computers, satellite dishes, modems, virtual reality, and artificial intelligence. But John O. Green sees the big picture and has fleshed it out with insights from the key players. A necessary prologue to the tale is his description of the networks' growth, from telegraph to telephone to high-speed computer. Bit by digital bit, each new technology enables a subsequent breakthrough. Today's Internet, for all its capability, is clearly but the precursor of something far more powerful.

One day our desktop computers will open out to the digital archives of our species. There is a word for the reviving of past knowledge through new forms of media—"renaissance"—and it is no overstatement to say that we have the necessary tools to make such an extravagant concept happen. Documents can be compressed and coded, transshipped through network hubs, and rebuilt on our desktops as needed. We're pushing toward the datasphere—a worldwide, networked digital library stuffed with multimedia text, sound, and video, accessible from home or office. It will take enlightened social policy to build and distribute such a resource, but the technology to do it is available today. A journey through these pages is a prologue to that future.

—Paul Gilster
Author, *The New Internet Navigator*

INTRODUCTION

Lo, soul, seest thou not God's purpose from the first?
The earth to be spann'd, connected by network,
The races, neighbors, to marry and be given in marriage,
The oceans to be cross'd, the distant brought near,
The lands to be welded together.
Walt Whitman, Passage to India, 1868

With the development of the Internet, and with the increasing pervasiveness of communication between networked computers, we are in the middle of the most transforming technological event since the capture of fire.
John Perry Barlow, 1995

What may emerge as the most important insight of the twenty-first century is that man was not designed to live at the speed of light.
Marshall McLuhan and Bruce Powers, Global Village

Progress was all right. Only it went on too long.
James Thurber, ca. 1950

ven if you believe only half of what you see, read, and hear, you may suspect that something VERY BIG is happening to us. It's usually called "the digital revolution" or "the new information age" or "the cyberspace frontier," and it's the subject of a current cascade of articles and books (including this one), talk shows, and TV specials.

Here's the refrain:

The revolution in information, communications, and media is well under way—some say crescendoing to its grand finale. This sea of change has been going on for decades. Depending on the analysis and/or the enthusiasm of the pundit, this is the most far-reaching development since—

    A. the microprocessor and the personal computer
    B. the Industrial Revolution
    C. the Renaissance
    D. the invention of the printing press and movable type

E. our forebears' career change from hunting and gathering to horticulture
F. the taming of fire.

The litany continues:
The revolution has now reached a stage where it permeates nearly every aspect of our lives. It will fundamentally and permanently alter the way we work and play, our family lives and our public ones. It is revolutionizing every aspect of commerce, local, national, and global politics, the art of war and the pursuit of peace, the fine arts, and micro and macro economics.

Now that *is* big. But is it big news or big hype? Or perhaps some of both? Is it possible that journalists, pundits, and prognosticators would do better to leave the evaluation of our moment to future historians who will be better placed to determine which, if any, of the above is true?

In any case, interesting things are afoot, and change and convergence are the watchwords. Some of the effects and affects we can predict, some we can only guess about, and some will arrive unheralded by anyone. The forces driving the digital revolution (if that's what we want to call it), in no particular order, are:

- *Moore's law, formulated in 1965 by Gordon Moore, one of the founders of Intel.* Really a prediction, it states that the power, speed, and capacity of microprocessors—the "brains" of computers—would double every 18 months and the cost would be halved in the same period. For 30 years that's held true, although how long it can continue is a source of some debate.
- *The new software that has co-evolved with this processing power.* Software engineers invent and refine programs that exploit increasing power and make computers continually easier to use (still not easy enough) and more useful, effective, and flexible. The power of the hardware and the sophistication and versatility of the software egg each other on.
- *The now fully fledged ability to digitize all forms of information and media (from text to voice to video), converting it into bits, the ones and zeros that are the lingua franca of all computers.* Once digitized, the bits can be compressed and sent winging through the pipes to be instantly reassembled at the other end in a perfectly faithful repro-

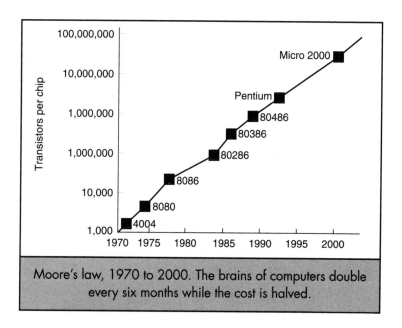

Moore's law, 1970 to 2000. The brains of computers double every six months while the cost is halved.

duction—or manipulated at workstations into strange, and sometimes wonderful, new media hybrids.

- *The rewiring with fiber optic cable and "wirelessing" of the planet—the so-called information superhighway.* We are rapidly increasing the bandwidth or "pipe size" of our global telecommunications network. Software engineers and telecom scientists are also inventing and evolving new data compression and transmission techniques for moving even more information and media more quickly and efficiently through the pipes or through the ether via satellite, microwave station, and cellular relay.
- *The surging growth of on-line communities on the Internet and World Wide Web and other networks.* These communities serve as the primary consumers, custodians, wizards, and special interest groups, promoting, protecting, maintaining, and innovating the new territories of cyberspace. The fact that there are millions of computers linked to the Net vastly increases the power and utility of any individual citizen's PC. What began as a specialized communication network for the academic research communities has become vast,

wide, and deep. Lay citizens, business and entertainment interests, and other commercial on-line services plug in and spawn on the Net. And the World Wide Web with its click-and-see interface makes it On-line for Everyone.

"Convergence" is the catchword, but the revolution is really the story of many convergings.

The most visible of convergences has been merger mania—now further stimulated by the Telecommunications Reform Act of 1996—as the big media companies, publishing houses, and software and computer companies buy or are bought by cable and network TV firms, the owners of communications satellites, power companies, and the telephone companies. The big players position themselves for the broadband wiring of the planet and for the delivery of the new programming and interactive capabilities, perhaps hoping for monopoly profits. But the new two-way structure challenges the old one-way delivery standard of broadcast media and print publishing. Once people discover the personal power and range of choice in new media and on-line services, they are not likely to give it up.

Another very interesting convergence is occurring in the area that business analysts, not generally given to poetry, call "content creation." The "creatives" are often finding unexpected bedfellows as writers, programmers, artists, graphic designers, film makers, and composers team up. These new production teams pool talents to explore the capabilities of this digital new world and create innovative multimedia melanges in entertainment, education, and art. Strange and wonderful hybrids emerge. Some are dismal failures in the market, and some make their creators millionaires.

From Wonder to Worry to So What

Reactions to this world wiring and the digital media explosion generally come in three flavors:

The credulous: "Oh! Wow! That's so cool!"
The cynical: "It's all a bunch of hype."
The alarmed: "This is out of control and dangerous."
The alienated: "I don't like computers. The future is passing me by."

Many people experience all of the above at various times. Along with the Utopian cheerleaders arise the pessimists and neo-Luddites: these critics feel that the digital revolution subverts our very humanity, and they express a variety of worries and denunciations on that topic—from intelligent to panicked and alienated in the extreme. Many others have a vague anxiety that they might miss the bus and be left behind.

There may be many things worth worrying about as we behold the digital future, but missing out is probably not one of them. The information revolution is constantly evolving to enfold all but the most disciplined technophobes.

One perennial school of thought says such-and-such (name your medium) is not good or bad in itself, it's what the medium is used for that makes it good or bad. Marshall McLuhan, the most famous of the media theorists, had a number of messages about media—including the admonishment that media *are* good and bad, in and of themselves, and that they exert their effects on our psyches and culture independent of the content that they carry. (Example: Regardless of what people watch, television has changed the nature of the family evening in America.)

One saving grace in the midst of the twentieth century's onslaught of new technologies has been our extraordinary ability to adapt to them. This ability is also our great predicament, but more on that later. We take to a new machine, enfold it into our lives, and quickly come into equilibrium with its benefits, demands, and detriments. Green's Axiom states that it takes no more than about 6 months between the initial sense of thrill, danger, setting foot on a new continent, and the flattening out of the learning and excitement curve to the point where we go for days or weeks taking the new capability for granted.

Into the Big Blue Yonder

My own curiosity about the digital age probably began decades ago when an engineer friend and acolyte in the computer priesthood of the 1960s took me on a tour of the campus computer center where the Big IBM was housed and ministered to. Carrying his stack of do-not-fold-spindle-or-mutilate punch cards, he ushered me into the presence of the blinking, whirring wonder and waited his turn at communion. The new machine had the aura of the Center of Power, Mystery, and Modernity as it efficiently gobbled up his programming cards, read them, and spat them out down the line.

ENIAC, the first large-scale electronic digital computer, was as big as a house. This behemoth ran from 1946 to 1955, when it was retired to the Smithsonian. Ever since, computers have been shrinking in size and growing in power.

No one could have convinced me then that in a few years I would have on my desk a mild beige console housing a machine more powerful and vastly more versatile and easy to use than the behemoth I'd met only a decade earlier.

I signed on for the duration in the early 1980s when I first grasped the implications of an early word processing program. As an aspiring writer who was also a miserable typist, I quickly saw the light. A few months later I had written and sold my first feature-length article, which arrived on the editor's desk blemish-free.

Victory of "the Blob"

The personal computer is our most concrete manifestation of Moore's law in action. Depending on your point of view (or mood that day) the PC can seem to be the wonder appliance of all time or the ultimate Trojan horse of end-of-millennium capitalism.

Insinuated onto our desks and into our homes, it feeds on pricey new software, and demands new hardware and software upgrades, a first modem and then a faster one, and occasional expensive repairs. If we want to stay in the game and keep that up-to-date edge, the computer requires regular replacement, like the family car, every 5 or so years for the latest wonder and ever more fully featured model. Over the past 15 years, the Protean general-purpose machine has been ingeniously programmed by software geniuses to subsume our Rolodex, our daily planners, our calculators, our record books, our financial transactions, our journals, our typewriters, and our reference books.

And now it has even gobbled up most of our communications devices: our telephone, fax machine, mailbox, and post office. For a growing number of people it's taking over as the conduit for newspapers, magazines, and television.

It's like the Blob of 1960s B science fiction movie fame, growing ever bigger on what it consumes, looming ever larger in our lives, demanding ever more of our workday time and attention. And not content to merely sit on the desk, the PC in stripped-down form of microprocessor has become an essential part of the innards of virtually every appliance and machine we deal with, from the refrigerator and washing machine to the automobile and touch-tone phone. Whether or not you own a PC, you're a computer user.

Truly, this silicon invader of our workaday and home life has (temporarily, at least) won a close to absolute victory. And now this subsumer of old media tantalizes us with new media vistas: the video phone, multimedia worlds, and fantastical new forms of communications, like multiplayer entertainments and virtual reality. The new means of melding text, video, graphics, and sound have only just begun to be explored and exploited by commerce.

At the end of the century, the millennium, the personal computer stands as our gateway and channel to the wired world and all its digital wonders and bafflements.

This Book's Mission

One way, perhaps the only way, to avoid being shocked by the future is to spend a bit of time there. Part of the purpose of this book is to offer readers a variety of windows on the future and as understandable as possible an overview of the "field" of telecommunications and new media.

No description of the future ever gets it right. The purpose of prophecy, envisioning, speculation and extrapolation, constructing utopias and dystopias is to imagine what the fruits of our strivings might be; what benisons we wish to attain; what dangers we must struggle to avoid. Imagining the future can also help us make some sense of the confusions of the present.

This book begins with scenarios of a future about 15 years from now. It then bounces back in time and reconstructs how communications and media have arrived where they are today—briefly citing contributions from a few of the key inventors, pioneers, thinkers, and visionaries of the nineteenth and twentieth centuries.

It then offers the reader a primer to the often off-putting and acronym-mined world of modern telecommunications. Next, it visits with the content creators: individuals and companies in the thick of developing new media products.

Finally, the book takes a look at the wildcard of artificial intelligence research and then takes an excursion into the "possible worlds" of the futurists and others who make their living by speculating about the shape of things to come. What are the implications of the new technologies for education, business, medicine, and entertainment? Where are we headed? What might be the social costs and benefits? Is it a win/lose game? If so, who is headed for the cat-bird seat and who is in jeopardy? How will we communicate with other humans in the near and longer-term future? And how will we communicate with these ubiquitous machines in our lives?

Along the way the book offers the perspectives of those who celebrate and promote the new technologies as well as those who worry about them. Wherever possible, the book attempts to separate hype, fantasy, and overblown expectations from the real promise and peril.

Optimists and Curmudgeons

This introduction began with four quotes reflecting disparate attitudes about technology and progress:

Walt Whitman, writing in the mid-nineteenth century, was among other things a technological optimist and democrat. This most American visionary poet felt that our technological destiny was bright and that the benefits would surely spread to the whole human family.

One line of his spiritual descendants came of age in the 1960s and 1970s and managed to combine visionary idealism with entrepreneurialism (also very American). Under banners like "Finally: A computer for the rest of us," some became quite wealthy inventing and marketing the machines and the programs to run them. Others have done well by spreading the technological news and excitement in an explosion of books, magazines, and on-line publications with names like *hotWIRED*, *CRYPT*, and *MONDO 2000*.

Marshall McLuhan, a Canadian professor of English and media theorist, wrote about the media revolution in a style by turns cryptic and brilliant. He saw both profound negative and positive effects in our technologies. Part of his message is: "It's here. It's real. We must understand it to deal successfully with it."

James Thurber's comment about progress represents that streak of Yankee irony that serves us well—so long as it doesn't devolve into an automatic cynicism about all change. Hopeful skepticism, to coin an oxymoron, might be a good way to approach the future in general, and cyberspace in particular.

Thurber's comment also reflects a deeper yearning of our times: "When are we going to arrive where it is we're headed?" (Or, as children ask at regular intervals on a long trip, "When are we going to get there, Dad?") When do we experience the end of all this change and dislocation and begin to enjoy the comfort of familiar surroundings? The apparent answer is "Not in our lifetimes."

We survive, and even thrive, with new technologies by co-evolving the personal, social, and cultural means to keep us afloat on a sea of change. The Scientific and the Industrial revolutions precipitated and enabled social inventions like the modern ideas of marriage, family, and childhood and political inventions like the nation state, participatory democracy, and an ever more inclusive idea of human rights and the value of the individual.

Nested just beyond the digital revolution and synergistic with it are others we can see, like the biotechnological and even the nanotechnological. There are probably some we can't see coming.

This book, then, is dedicated to us and our offspring, reluctant or enthusiastic "revolutionaries" who will wrestle with these technologies and use them to create new means for the human family to survive and thrive in a mutable world.

CHAPTER ONE

Scenarios with a Metabook
Anyone, Anything, Anytime, Anywhere

omewhere toward the end of the first decade of the next millennium, you'll be carrying around a small object about the size and weight of an appointment book—sometimes referred to as an omnibook, or metabook, or just "my book." When you open it, a screen unfolds that can display text, photography, and graphic design of a quality found in a "coffee table" book of our era. Small speakers produce sound and music that is crisp, clear, and high fidelity. Video looks and plays like film. Yours has a tiny video camera always aimed at your face as you look at the screen.

What you have access to, whether you are plugged in at your desk, sitting in your car, or walking in city or countryside, is astonishing, but you pretty much take it for granted as part of the fabric of daily life. It's a gateway through which you can reach anyone, anything, anytime, anywhere. All your friends, colleagues, and family and anyone else on the planet—either directly or through their messaging systems. Because of the clarity and a certain three-dimensional quality of the video and the sound, you almost feel that the person is present. Lovers actually do reach out toward each other and even touch the screen as they whisper the age-old terms of endearment.

If you're working on a project, or you simply have a particular interest you want to explore, you can also access all text, graphics, art, music, film, and video ever created in all of human history, as well as the wide variety of new media materials being created every day.

Although it sounds daunting to be "wired" to everyone and everything, you seldom think about it—no more than people of the late twentieth century felt in awe of the fact that they could place a phone call to any one of millions of people all over the planet.

One reason for your equanimity is that you have a software program, sometimes referred to as an "intelligent agent." You have been training your agent for years in the way you do things: your tastes, needs, preferences, and priorities. It also monitors everything you do and where you go on the Net and learns more.

In the morning when you click, touch, or ask for your account, the agent has collected a variety of world news headlines that anyone might see, the Big News, as well as headlines on new developments in stories or topics you flagged anywhere from a day or two ago to months ago. Touching or clicking on the head takes you to a brief synopsis. If you have time and want to go deeper, you can choose from two or three levels, each containing more detail than the last, as well as optional sound and video materials, including location shots and interview clips. At the deepest level you can access any and all of the full sources, such as the complete interview with a person who especially interests you, or all footage of a particular event or location, and a wide and deep range of related background text and other media sources. If you feel so inclined, you can instantly fire off a text, voice, or video message that may be viewed by the person you've been viewing and will definitely be viewed by his or her "agent."

You also have primed your agent with names of your favorite newscasters, journalists, commentators, and pundits, and the story defaults to their bylines and presentations. You can instantly mark any materials: text, video, or sound that you think you might want to return to—your agent knows exactly how and where to index these for you. If you're interrupted, the agent automatically marks your place.

Sometimes you come across a cartoon, an important piece of information, an unusual photo that you'd like to share with a friend or colleague. "Send . . . this . . . with . . . voicemail . . . to . . . Ellen . . . Silver." (Your book still has trouble understanding, if you speak normally andsluryourwordstogether. But your agent can recognize almost any single word you say, as well as many common phrases.) Your agent records your message and sends the packet to Ellen. Your agent knows many of your other preferences, right down to the percentage of time you like to see a talking head versus the source materials they're commenting on.

When a favorite film director, multimedia producer, poet, storyteller, singer/songwriter, interactive novelist, composer, or critic releases a new piece of work, your agent can notify you instantly or accumulate the information and hyperlinks for your later perusal.

In addition to the now ancient institution of e-mail, you also have the ability to record and send video mail messages, packet voice mail, or combinations of all three.

Kids and Whimsy

If your children allow you to view some of their to-and-from messages, you're astonished at the virtuosity they show. They're peppered with sound effects, zany clip art and photography, animations, and

Multimedia virtuosi of the 21st century?

whimsical writing. Simple digital tools for editing, changing, and manipulating media give everyone the ability to work quickly with sound, text, music, voice, video, art, photography, and graphics. Children who have been using these tools since they learned to read and write are astonishingly fluent with them. Both kids and adults collect over the years vast personal catalogs and source locators of their favorite writings and multimedia materials—all quickly accessible via intelligent agent.

A whole cultural phenomenon has sprung up around the art of multimedia signatures. Many people like to have a piece of music fade up at the end of their mail. People with the time and ability create their own. Others have even worked with a composer on-line to develop a personal theme. Musical signatures vary widely, from a piece of fiddle and banjo music to a blues riff to a grand orchestral flourish. Most people have a variety of musical motifs to be used according to the formality or informality of the piece of correspondence. Sometimes you send a voice mail message with music and a head shot of you or a landscape "beauty shot" or a montage of scenes from your recent mule pack trip in the Sierras. Or your very slick personal animated logo.

Over the years you've also tailored your interface with the computer to what's convenient to you. Depending on the situation, you turn on the touch-sensitive feature on your screen and point and drag with your finger or use mouse, track ball, voice commands, keyboards, and other pointing devices, alone and in combination.

All your work, play, and communications are accessible through file servers from wherever you happen to be, although many metabooks have an astonishing amount of onboard memory and processing power.

Business Unusual

The machine has changed the business world in interesting, unexpected ways.

The Meeting

There's a strong difference from company to company in how the ubiquity of video and multimedia affect the culture of meetings. All corporations still employ meetings in the flesh for ceremonial occasions,

Members of a work group chat with a colleague via impromptu video conference.

extremely crucial decisions, meetings of the board, and those occasions where "pressing the flesh," buttonholing, working the room, or just the intimacy of one on one are called for.

But the video meeting is fully rooted in the day-to-day function of work groups and project teams. Younger and older workers who have put in some thought and time in video communications workshops make good use of the medium.

People can tape themselves making a comment short or long, edit the piece, and correct the color and lighting. Some are more sensitive than others to issues of how they look and how convincing their "telepresence" is. (The quality of early video conferencing technologies were abominable, and many old-timers got turned off to it. But the new camera and screen technologies, the corrective software, the three-dimensional quality, and the dynamism of the protocols have made it a new ball game.)

For work groups who are separated in space (so many are now), video conferencing is a boon, and even people in the same building use it for ad hoc get togethers, short reports, quick questions, and consensus decisions. All video conferences can be digitally recorded, and some offices have it done automatically, unless someone opts to turn the record feature off.

Some work groups and extremely informal companies, like ad agencies, opt for a video conferencing package that is, well, quite special. Their video meetings have a very *dynamic* character. The screen frame of the person who is speaking increases in size while he or she talks. People who want to speak have a light or "hand-raised" icon by their frames. If someone interrupts, his frame begins to swell, while the interruptee's shrinks. This can go back and forth while the exchange takes place, until the new speaker is established. If someone who's been interrupted is dying to reply but is "holding his tongue," his screen pulses indicate variable levels of impatience.

At the ad agency, if you have a problem with what someone's saying, you can draw a diagonal red stripe across his face, or morph (blend) it with the head of a rhinoceros, orangutan, or a particularly pedantic U.S. Senator—whatever you have on hand in your "mockery" file. Naturally, these meetings can get quite lively and can even get out of hand. There are sometimes meetings about the meetings. One night, late, the creative director strips the software from everyone's system except hers. But a graphic designer serendipitously has made a copy and has it back up for everyone next week, when things have cooled down.

Needless to say, very few banks or insurance companies use this sort of volatile and rude conferencing software, although some of the dynamic possibilities of new video conferencing make a restrained appearance.

New Media Arts

Recently, the whole world—or so it seems—has been buzzing about an extraordinary multimedia composition by a young media wizard who first started playing with multimedia on her parents' AV Mac in 1995 when she was only three years old.

The whole piece is only 42 minutes long, but it's had an extraordinary affect on people. It premiered to a large audience before she put it on-line. At points in the composition the whole audience or the individual can interact with the piece. A storm of controversy has built up around it. In fact, at the premiere some people walked out, most were cheering, some were in tears, and others argued tensely in groups.

The controversy is over a new technology developed by the composer with her friends and collaborators that has enabled her to take old video and sound of people dead or living and construct completely new, utterly lifelike scenes—that never happened.

In one scene a deceased president, a notorious terrorist, and the composer's dead parents meet beside a river and sing a song together. The ability to construct such a scene is the culmination of a technology first hinted at in a film from the 1990s called *Forest Gump* in which the lead character shares the stage with famous figures of the recent past.

Only this is so *lifelike*. There is no way to tell that the people are not real, or that the event or scene is a complete fabrication. Some critics are outraged and call the scenes monstrous and ghoulish. Others focus on the whole piece for its undisputed beauty and power. Millions, even those disturbed by parts of it, have favorite scenes they play over many times.

Others are worrying publicly about this technology getting into the hands of criminals, blackmailers, con artists, and the like—even though the composer explains that the scenes took hundreds of hours to appear lifelike.

In another startling and disturbing part of the piece, you glimpse yourself among the passengers on an airliner just before it crashes.

And at the end—the last 4½ minutes—you revisit characters you chose to follow, images you'd responded to earlier, and music that captured you. It's all reprised in a finale that has unlimited variations, according to how you participate—and by some miracle of the composer's art, all of them seem somehow right and powerful.

Teachers, Students, and the Global School

Although problems of resources, infrastructure, and haves and have nots still exist, the new connection technology has brought at least two major benefits to education. Virtually all schoolchildren have their own metabook. Even the poorest have an inexpensive version provided through private and public foundations. The cheaper versions, sometimes called webtops, are purely connection devices to the World Wide Web and other networks, and the software they use is from a base of public domain shareware. The connectivity makes even the cheapest sets extremely powerful by today's PC standards.

Paradoxically, the ubiquity of metabooks has taken away some of the anxiety and former hysteria about "computer education." The idea that children needed to be taught computer literacy is as foreign a notion to teachers and parents as the idea in our time would be that children must be taught how to open a book. It is widely accepted, though, that children need to learn through real world, hands-on experiences, away from the classroom if possible. In the early grades, especially, use of metabooks is minimized.

The Web and universal connectivity have proven a great boon to

The World Wide Web is a great boon to teachers and financially strapped school districts.

teachers and financially strapped school districts. Over the past decade educational foundations have sponsored a wide and growing range of superb curricular multimedia materials in the sciences and humanities for all grade levels. These are available free anytime to all teachers and students—and anyone else, for that matter. The materials provide teachers with everything from highly structured "ready to go" lesson plans to as much choice in developing their own teaching materials as they have time and energy for. Students who develop particular interests in a topic can explore as deeply as they want through hyperlinks to primary sources of text and multimedia.

Another important development has been the phenomenon of "student-teacher explorer teams." Often sponsored by corporations, these teams have fanned out over the globe—and beyond. Students and teachers in the classroom can go on-line with the teams wherever they are: at undersea exploration sites, astronomical observatories, ecological and engineering projects, archeological digs—the list goes on and on. During on-line visits, the teams discuss the science and latest developments or observe ongoing operations with commentary from the field teams. Students in the classrooms suggest experiments to their counterparts in the field. In fact, senior scientists and principal investigators have been astonished more than a few times by important insights and discoveries made by students through these classroom partnerships.

One of the most popular on-line sites is the student-teacher explorer team on the space station. Students have devised a wide variety of physics, biology, and materials science experiments (as well as art and dance projects) that range from silly to ingenious. With the enthusiastic comments and collaboration of their earthbound counterparts, the "space team" has invented and refined a variety of games that can be played only in zero gravity.

Students are nearly always surprised when they learn that their taken-for-granted connection capabilities were mostly unavailable to students in the last century.

CHAPTER TWO

Something Is Bound to Come of It

It often happens, with regard to new inventions, that one part of the general public finds them useless and another part considers them to be impossible. When it becomes clear that the possibility and the usefulness can no longer be denied, most agree that the whole thing was fairly easy to discover and that they knew about it all along.

Abraham Edelcrantz, inventor of the Swedish optical telegraph

o understand modern communications and media and where they're headed, it helps to look back at several independent tracks that are now all converging, through some kind of interesting trick of synchronicity, precisely in time for the millennium. This chapter focuses in particular on two critical strands: the evolution of distance communications and the rise of the microprocessor and general-purpose computer.

The Engine of War

Many of the early developments in communications, like so many other technologies, owe their genesis to the arts and exigencies of warfare. Commanders have communicated with their officers and troops with gongs, bells, pipes, drums, rattles, horns, cannon, and rifle fire. Visual media have included smoke, flags, banners, hand and arm signals, semaphore and, at night, torches, flares, and rockets. Runners, riders, carrier pigeons, and trained dogs have all been used. Genghis Khan directed his forces to attack by firing whistling arrows toward the enemy host. The

Carrier pigeons as used in ancient Syria, a woodcut from Jean de Mandeville's *Travels in the Orient*, published in 1481. Homing pigeons were valued as message deliverers before the days of modern communications.

ancient Gauls erected stone towers from which they warned each other about the encroachment of Caesar's armies and other invaders by a system of shouts and cries pitched to carry from tower to tower.

The First National Network

In 1792, Claude Chappe developed the optical telegraph, an invention that incorporated many of the features of the modern communications system.

Frustrated in his efforts to make an electrical telegraph, Chappe instead developed a mechanical device that sent visual signals. In a trial run, a doctor friend sent Chappe the first message: *"Si vous réussissez vous serez bientôt couvert de gloire."* ("If you succeed, you will soon bask in glory.") The whole message took about 4 minutes to transmit. Chappe had first thought to name his invention the *tachygraphe*, or "fast writer," but it was dubbed more appropriately *télégraphe*, or "far writer."

Chappe soon evolved the transmitter design into a device with large pointers that could be spun on arms attached to a central post called the regulator. The whole contraption sat on top of a stone tower, perhaps harkening back to Gallic forbears. By varying positions of the pointers in 45-degree increments, the signal master could transmit hundreds of symbols. In the process of implementing the system Chappe encountered and solved a number of modern network problems, including data compaction, error recovery, flow control, and encryption.

In 1794, a line of Chappe's telegraph stations connected Paris and Lille. Napoleon quickly grasped the value of the telegraph for military communications and rapidly expanded the system of towers and signaling arms throughout France. He even took portable units on the disastrous Russian campaign. By 1844, the network connected 29 cities through some 500 stations over 4,800 kilometers and employed 3,000 operators and workers.

The Electric Age

No, the electric telegraph is not a sound invention. It will always be at the mercy of the slightest disruption, wild youths, drunkards, bums, etc.
  A French media critic, 1846

The French were quite proud of their system and kept it in place years after electric telegraphy became a proven substitute. But the vast distances and land mass of North America required a faster technology, and the dawn of the digital age actually arrived some 150 years ago with the

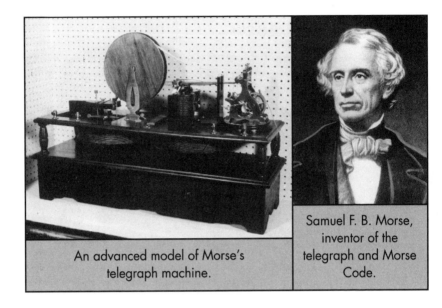

An advanced model of Morse's telegraph machine.

Samuel F. B. Morse, inventor of the telegraph and Morse Code.

first electric wire network. The "dits" and the "dahs" of early telegraphy were digital communications formed by opening and closing a circuit.

Although it was clear as early as 1816 that an electric telegraph was feasible, it wasn't until 1840 that an American, Samuel Morse, invented a practical receiver for the telegraph by harnessing the recent invention of the electromagnet. He also invented Morse Code, which defined patterns of dots and dashes to represent letters and numbers used in the messaging. In 1844 Morse inaugurated the first public telegraph line with the message "What hath God wrought?" (The refrain resounds ever since with each new electronic wonder.) At first, he used a pencil stylus above a moving tape to record the dots and dashes—to be decoded later. This early system, which presaged later recording devices, was abandoned as human operators learned to quickly decode in "real time" the clicks of the instrument.

The drive toward greater speed, higher capacity, and more flexibility in communications had already taken hold. Morse and others began looking for the means to send two messages from opposite ends of the same wire, called duplexing, which would be essential for the networking capacity and the business development of the telegraph. They solved this problem, and the telegraph network in America grew rapidly. By 1876, it included 214,000 miles of wire and 8,500 telegraph offices.

Then I Put My Ear to a Bell Tellyphone

Alexander Graham Bell's invention of the telephone arose from a lifelong personal interest in the phenomena of speech and hearing shared and encouraged by his whole family. His mother was deaf, and his father and grandfather had trained him in the art of communicating with the deaf and also public speaking. His interest in the physiology of hearing helped him conceive and create his invention; the public speaking helped him demonstrate and promote it, and throughout the years effectively defend it and the fledgling Bell system against other patent claimants and rival entrepreneurs.

Bell combined his interests in speech and hearing with a tinkerer's fascination with electricity. He and his brother rigged up many bizarre, even macabre, experiments, which often involved squeaking, groaning, and squawking kludges of electromagnets acting on metallic diaphragms. Their efforts also included attaching artificial vocal cords in a human skull and operating them with compressed air in a way that purportedly lured over a neighbor woman with its repeated rendition of the word "mama." Once, they managed to obtain an ear from a cadaver and attached a straw to the small bones of the inner ear. When someone spoke the straw traced sound waves on a glass plate.

Bell became a professor of speech at Boston University and focused his experiments on developing the telephone. Finally, in 1876, he and his assistant, Thomas Watson, a skilled mechanical engineer and craftsman, refined the design so that the latter was able to hear the first faint words uttered over a phone line: "Mr. Watson, come here, I want to see you."

Bell filed his patent a mere 2 hours before Elisha Gray, a professional inventor allied with Western Union Telegraph, filed a very similar design. Gray had sketched his out months before but had neglected to patent it because he felt it was a toy, a curiosity with no commercial value. This opinion became his and Western Union's official line when asked about the future of telephony. The "toy" tag is still a frequently heard refrain of newness envy at the unveiling of a new communications technology.

The mechanism of the early phone worked roughly like this: Sound waves from the speaker's voice vibrate a metal diaphragm, which in turn

The first New York to Chicago telephone line was opened by Alexander Graham Bell in 1892. When Bell died in 1922, phones were silent for two minutes.

compresses a layer of carbon granules, which in turn create a varying voltage in an electrical current in the phone line. At the other end the process is reversed, and the variable electric current gets transformed back into sound waves that create a replica of the speaker's voice. The genius of the telephone is not only that it allows you to "call" to someone over a great distance without some superhumanly loud shout. It also vastly speeds up the transmission of the sound from the pokey speed of sound through air to the rapid clip of electrons in copper (and now light in glass fiber). A "call" through the air from New York to San Francisco would take about 4 hours to get there. Through wire, of course, it's nearly instantaneous. (And you don't have to shout.)

Curiously, Bell himself wasn't quite clear on the real future for the telephone. He at first imagined it being used to create a kind of cable radio, sending news, music, and other entertainments. In an early demonstration, Bell alarmed and fascinated theater audiences with a presentation that included an organ performance by Watson, who was playing across town, that was transmitted through a wire to a powerful receiver amplifier in the theater.

As Bell began to attract backers, they gravitated toward the idea of person-to-person vocal communications via a networked system, and the Bell System was born. Thomas Edison was contemporaneously inventing the new media of electric light, the phonograph, and moving pictures. As with the light bulb and the lighting system, it was the phone *system* that turned the toy into the most powerful communications network of the twentieth century.

Hackers in Knickers

In his excellent book *The Hacker Crackdown*, author Bruce Sterling provides a captivating account of Bell and the early years of the Bell System and AT&T.

He reports on one of the earliest sightings of the hacker phenomenon in 1878, the first year of the Bell System. At the telephone office, teenage boys (with the usual low-boredom threshold and surging hormones) employed as clerks, operators, and janitors made nuisances of themselves by mouthing off to customers, concocting clever sabotage of the switchboard, and disconnecting callers or plugging them into the wrong number. They'd discovered, as modern hacker/crackers have, that the anonymity of the network enabled them to commit mischief and mayhem as the spirit came over them—which it frequently did. The boys soon got the boot, and the company replaced them with young women operators, who became the human emblem and voice of the system for decades.

Another interesting phenomenon was the effect of cultural and political differences on how a country viewed and implemented phone technology. In Hungary, the government set up Radio Hermondo along the lines of Bell's early conception: a sort of telephone/radio system. At certain hours

Switchboard operators, 1895.

of the day the phone would ring, owners would plug in a trumpet loudspeaker, and would then receive music and government news of the day.

The British tended, at first, to see the telephone as a vulgar intrusion on privacy, which allowed any fool into your household at any time of day. If you had a phone at all, you tended to keep it in a closet, hallway, or some other out-of-the way part of the house, such as the servants' quarters.

Stalin perceived general telephone service as a potentially dangerous subversion of central authority. And when it did arrive the system was unreliable and heavily tapped. Even into the late 1980s, in the Soviet Union, when you wanted to place a call to another country you made an appointment with the operator (presumably to allow time to set up the tap), and she would (or would not) call you back with the connection at the appointed hour.

The French had their own suspicions about the telephone as an instrument of hooliganism and uncontrollable outcomes, and for decades their system's trademark was its "snarled inadequacy."

In America, although the early adopters of the phone (as with subsequent communications technologies) were the rich, professionals, and technical hobbyists, the phone quickly developed a democratic character and appeal. Theodore Vail, the first president of AT&T in 1885, created a corporate slogan that encapsulated his and Bell's own vision for the telephone: "One Policy, One System, Universal Service." Universal telephone access and service became the company's goal for the future—another instance of democratic ideals also being good business. The system spread through New England and then rapidly through the rest of the United States. By the mid-1920s it had become possible to make a connection between any two places in the country.

A Father of New Media

> *Presumably man's spirit should be elevated if he can better review his shady past and analyze more completely and objectively his present problems. He has built a civilization so complex that he needs to mechanize his record more fully if he is to push his experiment to its logical conclusion and not merely become bogged down part way there by overtaxing his limited memory.*
> Vannevar Bush, 1945

> *Vannevar Bush probably had more influence on the growth of 20th-century science and engineering than any other single human being.*
> Bruce Sterling, 1995

In 1945, Vannevar Bush, then director of the Office of Scientific Research and Development for the U.S. war effort, published a feature article in the *Atlantic Monthly* entitled "As We May Think," which has become, over the past decade, one of the most cited pieces of science writing ever. (It even has its own page on the World Wide Web, complete with full text of the article and a photo of the charismatic-looking author.) Bush begins with the question: What are the scientists to do next? The piece then goes on to anticipate the fax machine, xerography, the Polaroid camera, microfilm, speech recognition, and the long-heralded speech-to-print typewriter, the personal computer, multimedia, and hypermedia. The piece even presents a tantalizing augury of virtual reality.

It has become one of the hallmarks of twentieth century warfare that the opposing sides must harness the efforts of scientists and engineers to roll out and "force-converge" technologies throughout the R&D and testing phase and into production as quickly as possible, and before the adversary. The game is to one-up and "leapfrog" the enemy's newest technological capabilities, or better yet, develop weapons, communications, and intelligence-gathering systems that the enemy hasn't even thought of yet.

Vannevar Bush, circa 1930.

One feature of the cold war was that it sustained a warfare mode of R&D and kept money flowing to it, which has much to do with the surge in technology in the last half of our century. But in 1945, Dr. Bush had a different agenda. His article is, in part, a call to the science, engineering, and business communities to continue their teamwork to create an ongoing peace dividend of constructive and useful technologies for the postwar era. He had been witness to, and instrumental in, the creation of "miracles" used to vanquish an enemy and win a war.

He wrote:

> The applications of science have built man a well-supplied house, and are teaching him to live healthily therein. They have enabled him to throw masses of people against one another with cruel weapons. They may yet allow him truly to encompass the great record and to grow in the wisdom of race experience. He may perish in conflict before he learns to wield that record for his true good. Yet, in the application of science to the needs and desires of man, it would seem to be a singularly unfortunate stage at which to terminate the process, or to lose hope as to the outcome.

Bush's desire to "encompass the great record" rose in part from his keenly felt frustrations at the dawn of the information explosion. No single scientist could keep up with all the new developments—even in one's own field. Bush's article describes a hypothetical machine called a Memex extrapolated from technologies Bush had helped birth during the war. The Memex would provide new and powerful ways to keep up with the info crunch by enabling the user to store and quickly retrieve large amounts of textual and photographic materials.

These visionary challenges inspired two generations of computer scientists who have now endowed the modern multimedia PC with nearly all the information, document handling, reproduction, and communications capabilities that Bush yearned for.

Acceleration into the Electric Age

The world has arrived at an age of cheap complex devices of great reliability; and something is bound to come of it.

Vannevar Bush, 1945

The postwar years were boom years for new developments in mass media, telecommunications, computer technology, and the further wiring of the United States and the planet.

During the war scientists built the computer ENIAC primarily to calculate ballistic trajectories. ENIAC was a true monster, filling a whole room and reportedly dimming the lights in Philadelphia whenever the scientists turned it on, although it had only slightly more processing power and "intelligence" than a hand-held calculator. It also used thousands of vacuum tubes, which burned out at frequent intervals, requiring a team of technicians to race around the room replacing them.

In 1948, a team of engineers at Bell Laboratories—Walter Brattain, John Bardeen, and William Shockley—invented the transistor. It performed the same functions of signal amplification and circuit switching of the vacuum tube. But because of its far greater durability, small size, and low heat production, it was to become the king pin in the new era of miniaturized electronics, integrated circuits, and microprocessors.

But television was the biggest story of all. In 1941 came the first regularly scheduled broadcasts (a few years after England and France) from the Empire State Building. In 1945, Americans owned a very modest number of TV's: 5,000. By 1962 the number had exploded to 59 million.

The TV has had a much more complex and contested parentage than the telephone. If you are British, you might attribute it to John Logie Baird,

Philo T. Farnsworth with his newly developed television device.

who as early as the 1930s was using the BBC's radio transmitters after hours to send his electromechanically generated images to subscribers. If you're American, you might attribute it to Philo T. Farnsworth, who, after transmitting an image of a dollar sign (perhaps as an encouragement to his backers), filed for a patent for the first all-electronic TV system in 1927. Or you might lean toward Vladimir Zworykin of RCA, who filed a patent for the first crude electronic TV camera in 1923, and then patented many improvements over the next few decades. Farnsworth's claim received validation in 1939 when, after years of litigation, RCA was required to pay him royalties.

Germans, though, will insist that the parentage goes back even further, to 1884, with Paul Nipkow's invention of what he called the electric telescope. The system used a spinning disk punched with holes that captured portions of the image of a moving object and converted them into electrical signals reconverted by the receiver into neon light and projected by a similarly spinning disc onto a groundglass screen. It was not something we would recognize as television, but it established the principle that to capture and transmit a moving picture the picture would have to be broken down into discrete lines and frames (one complete picture), reassembled at the receiving end, and projected at a minimum of 24 frames per second to create the illusion of unbroken motion. In fact, the European format eventually settled at 25 frames per second at 625 lines per frame, with the American NTSC standard at 30 frames per second at 525 lines per frame. (Naturally, the two formats are completely incompatible.)

Vladimir Zworykin with an early RCA television tube.

America's Charles Francis Jenkins and Britain's John Logie Baird both developed commercial TV systems based on improvements to Nipkow's invention, though the low-resolution and flicker sometimes caused headaches in their viewers. British electrical engineer A. Campbell Swinton, like Leonardo sketching flying machines he could never build, laid out in 1908 a complete scheme for a modern television camera and receiver based on the cathode ray tube. He described a device that rapidly scans an image electronically line-by-line and reassembles it in the receiver's cathode ray tube using an electron gun surrounded by magnetic coils to sweep a variable electron beam back and forth across the screen. Farnsworth, Zworykin, and others used Swinton's blueprint as they began their race to father television.

Forty years and $50 million dollars later, RCA finally had a system that was ready for prime time.

The New Village Messenger

I invite you to sit down in front of your television set when your station goes on the air, and stay there. You will see a vast wasteland—a procession of game shows, violence, audience participation shows, formula comedies about totally unbelievable families . . . blood and thunder . . . mayhem, violence, sadism, murder . . . private eyes, more violence, and cartoons . . . and, endlessly, commercials—many screaming, cajoling and offending.

> Newton Minow, Chairman of the Federal Communications Commission, in a speech to the National Association of Broadcasters, 1961

. . . the stakes are very high, and the need to understand the effects of the extensions of man becomes more urgent by the hour.

. . . in operational and practical fact, the medium is the message. This is merely to say that the personal and social consequences of any medium—that is, of any extension of ourselves—result from the new scale that is introduced into our affairs by each extension of ourselves, or by any new technology.

> Marshall McLuhan, Understanding Media, 1964

What was going on? Where were we headed? And what in the world did this electronic age mean to humans and the human family? Was networkedness a blessing or a monster in the disguise of good, clean American commerce? It was time to do some reflecting, cogitating, assessing, and forecasting. And in the mid-1960s Marshall McLuhan arrived on the scene, having done plenty of all of the above. McLuhan—a professor of English in the United States during the late 1930s and early 1940s and afterwards in Canada, and an inventor of media studies as a college subject—delivered to a fascinated new audience of intelligentsia and media movers and shakers an outpouring of sometimes cryptic, often penetrating, and generally nonobvious ideas about media in general and TV in particular, the hot button of the times.

He had been trained in literature at Cambridge, eventually earning his Ph.D. there in 1943, but saw the literary tradition spawned by the written and printed word coming to the end of its reign. He saw it being supplanted by what he called "the electric age," and he could see powerful effects both good and bad in the transformation.

He wrote in *Understanding Media: The Extensions of Man:*

> The electric technology is within the gates, and we are numb, deaf, blind and mute about its encounter with the Gutenberg technology, on and through which the American way of life was formed. It is however no time to suggest strategies when the threat has not even been acknowledged to exist. I am in the position of Louis Pasteur telling doctors that their greatest enemy was quite invisible, and quite unrecognized by them. Our conventional response to all media, namely that it is how they are used that counts, is the numb stance of the technological idiot. For the "content" of a medium is like the juicy piece of meat carried by the burglar to distract the watchdog of the mind.

Marshall McLuhan, writer, educator, and pioneer communications theorist, in 1966.

With the publication of *Understanding Media* in 1964, McLuhan became the oracle to consult and argue about if you wanted to understand the effects of media shock. He became famous for the phrase "the medium is the message"—an utterance only slightly more communicative than "$E = MC^2$." But it was very likely a conscious attempt on his part to compress all of his scholarship and thinking about technology, media, culture, and change into something as compact and catchy—in a recondite sort of way—as Einstein's famous formula.

Part of what it means is this: The nature of the technology and its diffusion is more important than any particular transmission or set of transmissions. The effects of the medium, for good and bad, on culture are quite independent of the contents of that medium. And the changes we're undergoing as individuals/community/society are hidden from us because we are so involved (hypnotized, McLuhan would say) in the contents of the conversation we're in, the book we're reading, the TV we're watching, or the computer we're swearing at. Those senses that are not engaged by a particular medium are somnolent.

McLuhan was not necessarily nostalgic for the fading reign of print. He felt that in addition to the wisdom, truth, and beauty it had conveyed, it also fostered negatives, like bureaucracy and an artificial preference for linear thinking and a one-dimensional idea of rationality.

One wonders what he would have had to say about the digital revolution, high-definition television, multimedia computers, and the highly nonlinear worlds of on-line cyberspace.

Making the New Machine Ready for New Media

As man succeeds in translating his central nervous system into electronic circuitry, he stands on the threshold of outering his consciousness into the computer.

Marshall McLuhan, *Global Village*

What properties do [computers] have that are so special and powerful that the shape of society will once again be changed? The first property is that all previous marking and symbol systems can be subsumed by digital representations. The implications of this become really interesting when the second property is added: that stable digital representations can be inexpensively made in astronomical quantities. Third, the representations can be transmitted to everyone in the world very rapidly and at low cost. Finally, the digital representations are active and reflective; they can read and write themselves at great rates of speed.

Alan Kay, co-creator of the Alto computer

Bill Gates and the two Steves, Jobs and Wozniak, are household names known for their visionary entrepreneurship in the computer world, but as is often the case, behind legends are the "legends to the legends," who offered giant shoulders to stand upon. In the case of the personal computer, these were engineer dreamers and pioneers from the new field of computer science and the even younger field of artificial intelligence. Even though the machines they'd teethed on in those days were the huge, expensive tools of the computer priesthood, lodged securely in the government, military, and megacorp, these pioneers believed that computers would change the world by becoming accessible to everyone. They realized that for this to happen normal, nontechnical people, even children, would need to be able to "communicate" easily with the computer. This was not considered obvious or even generally desirable at the time.

The Crusader

In 1945, Douglas Engelbart was a young naval radar technician waiting to come home from the war in the Pacific when he read Vannevar Bush's *Atlantic* article. It moved him profoundly.

Steven Levy in his book *Insanely Great*, on the genesis of the Macintosh, describes how, over time, Engelbart developed a sense that he was on a crusade to create the tools and the means for humans to augment their natural abilities and to vastly improve the ways they communicate and solve problems together.

Driven by his sense of crusade, he left a perfectly good job with NASA and earned his Ph.D. in the fledgling field of computer science, got a small government grant, and set up the Augmentation Research Center at Stanford Research Institute (now called SRI International). With Bush as his inspiration and a brand new knowledge of what computers could do—as well as a vision of what they should be made to do—he published a paper in 1963, "The Conceptual Framework for the Augmentation of Man's Intellect." In it he predicted the advent of personal workstations networked together, anticipated such utilities as the on-line dictionary, and described the key power the word processor would have as a cutting and pasting combinatory engine. He wrote, "Trial drafts can be rapidly composed from rearranged excerpts of old drafts . . . you can integrate your new ideas more easily and thus harness your creativity more continuously." All of this was at a time when the "interface" for most computers was still a stack of punch cards. (The interface is, of course, the ways and means by which the human interacts with the computer.)

Although the paper attracted little comment at the time, the Augmentation group continued its work and soon spawned two of the fundamental interface technologies of modern computing. Their first innovation was the window, which became the central screen metaphor of the Macintosh interface—and now, of course, nearly all other computers through Microsoft's famous software version—Windows.

Windows allow you to "see into the computer" and what's stored there. To an important extent it makes the machine transparent, at least in terms of any document, application, graphic, photo—or anything else you work with in your machine. It also makes your own work transparent and accessible to you.

As Steven Levy writes, "Windows are really quite profound. Using them implicitly reshapes our relationship to information itself . . . a digital peep show where we flick open the shutters to information."

It's as though you could reach into some ghostly file cabinet, grab a document, and place it anywhere on your desk—all without disturbing the file cabinet or its contents. By a neat trick of cyberspace you are simultaneously looking at documents, changing them, combining them, and sending copies of them off to someone else, when all the while the documents are also sitting safe and undisturbed where they are stored. Soon enough a document would be able to contain not merely text, but all other communications media. The window metaphor turned out to be an essential step toward realization of Bush's Memex dream.

Thirty years later, we take windows for granted. We even become annoyed if our computer runs out of memory and we have to close an application window before we can open another, wasting precious seconds before we can get back to work. Never mind that not so long ago accomplishing the same change would have involved getting up and going to a file cabinet, searching for, and, if lucky, finding the right file. Then we'd return to the desk, or perhaps move over to a drafting table or a dark room, finding and preparing all the appropriate tools and workspace for the next task, and so on. But now, we want it all instanter. Green's axiom is in effect. The computer has changed all our expectations of what it takes to perform a particular task.

The Mouse That Scored

Having digits (curiously the root of "digital"), it's natural for us to point in communicating. We select, we motion "put that one here." We thumb through papers and poke around; we "make a point." We "point out," "point the way," and even mark points in time. So a digital pointer makes all the sense in the world. But back in the mid-1960s, it was a peculiar novelty, or worse, a nuisance and a toy no serious computer professional, or propeller head, would want to use to operate his system. Nowadays, of course, most computer users would be stymied without one. It's the essential equipment (and only device needed) for navigating the Web and the "journey of a thousand clicks."

The mouse and the track ball (sometimes called an upside-down or "dead" mouse) or some other pointing tool are basic to the modern interface. And Engelbart and company invented it—and gave it all

Engelbart's early wooden "mouse."

Douglas Engelbart.

away. At a legendary event in 1968, Engelbart demonstrated his "office of the future," boggling his fellow computer scientists with the new look and functionality of the windows computer and the navigational prowess of the mouse. His computer was also (courtesy of AT&T) networked miles away to his work group, which at one point took over control of the demo. He had set a benchmark for the human-machine interface.

But the government giveth and taketh away, and Engelbart lost his funding and his group—if not his dream. Others, though, were already reaching for the magical baton he was passing on.

The Alto: Making Computing Safe for New Media

The great computer science institutions and artificial intelligence (AI) labs of the 1960s, among them MIT, Carnegie-Mellon, and Stanford, were graduating a new generation of talented and ambitious scientists who dispersed throughout the country to government, university, and corporate research posts. All of them had met or at least heard of each other and each other's work and best ideas.

One of the greatest pools of computer science talent in the early

The Alto computer, out of the "dark."

1970s ended up at Xerox's Palo Alto Research Center (PARC) with a mission to take the computer to a new level of human compatibility.

The campus was nestled in the foothills of the San Francisco peninsula, then and even now—after the Silicon Valley revolution—some of the loveliest real estate on the planet. It was an ideal human space and the perfect place to tease the computer to a new level, a more human mode of communication—to put the "face" in interface.

A brilliant young programmer and composer, Alan Kay, was one of the senior scientists on the project. Like Engelbart and other computer science visionaries, Kay had also had an epiphany about the destiny of computers and society. It came while he was reading McLuhan's *Understanding Media* and grasping the meaning of "the medium is the message." Kay writes of realizing: "The computer is a medium! I had always thought of it as a tool, perhaps a vehicle—a much weaker conception. What McLuhan was saying is that if the personal computer is truly a new medium, then the very use of it would actually change the thought patterns of an entire civilization."

Kay had learned many lessons from an earlier, unsuccessful attempt to design "a personal computer." He also had developed new goals for the hardware and software he wanted to design. Even a child, especially a child, should be able to use it and harness it as an intimate creative tool. He wanted to create a pleasurable "user illusion," keeping in mind always that "what is presented to one's senses *is* one's computer."

And what Kay and the wizards of PARC developed literally brought computers out of the dark ages.

While everyone else (those few who had access to computers) worked with light characters on a dark screen, responding to ambiguous screen prompts like *A:?*, the Alto, as the PARC computer came to be called, presented dark characters on a pleasantly lighted screen, like the printed page. When you walked into an office at PARC you felt that you were already in the future, that it all worked and the lights were *on*. Visiting a computer lab of the old sort, one had the impression that people were staring into tunnels or down rabbit holes.

After two years of development, from 1972 to 1974, this is what the PARC team had produced:

- The Alto functioned with a friendly new programming language called Smalltalk.
- The machine furthered the "desktop" metaphor Engelbart had opened with his windows. And, it, of course incorporated his mouse as the pointing tool.
- The team got rid of the old "mode" metaphor wherein you had to be in one mode to edit text and yet another to make a copy of a file or diskette—all the while consulting your command card for the cryptic enabling commands.
- The window, in the PARC team's scheme, was the active file and the current mode—all accomplished by moving the cursor and clicking the mouse.
- They had done wonders with the text editor.

With early word processing programs, moving or deleting a paragraph or block of text required that you first be in the right mode for such an operation. You didn't just "do it." You had to insert the cursor at the beginning and invoke one block command and then insert it at the end and invoke another. Then having, of course, moved your cursor to where you wanted the block to appear, you had to invoke the move command, and the deed was finally done. Even this klutzy routine was a boon for writers because, for all its awkwardness, it meant never having to retype perfectly good text just because you needed to do some radical surgery on a page or paragraph.

- Most important, the bit-mapped, high-resolution screen, combined with programmed and programmable "objects," allowed the modern graphical user interface (GUI) with its desktop characteristics. The bit-mapped screen was an extremely expensive feature at the time, the memory alone costing some $7,000 (the same amount of memory now costs less than $100), but it was absolutely crucial.

Bit-mapping assigns every area of the screen an "address." Think of the screen as a large piece of graph paper. Each square is a pixel, short for "picture element." On a high-resolution screen the pixels are tinier and there are more of them (the Alto's screen had 500,000), but each can have an x and y axis address all described in bits. The computer can thus keep track of the position of any graphic and word, as well as the current location of the pointer being moved around by the mouse.

The bit-mapped screen also allowed the introduction of graphical objects with programmed characteristics, called icons. An icon of a file folder can be named, can be filled with files and other folders containing files, and can be opened and closed. It can be moved around the screen with all its "contents" intact, and it can be copied or thrown away.

Another crucial feature of the PARC system was that all Altos in the facility were networked together via the team's invention of Ethernet, which enabled team members to communicate and collaborate.

Although Xerox correctly dubbed the system "the office of the future," the company never quite knew what to do with all the extraordinary innovations coming out of its research group—but others did. Two glitter-eyed entrepreneurs named Gates and Jobs had toured the facility separately and been given demonstrations of the Alto. They saw the light at PARC and knew that it was good.

CHAPTER THREE

How the World Is Wired

Watching Fire

Pundits and info gurus have compared the significance of the telecommunications revolution to the capture of fire. It's useful, though, to understand and imagine the taming of fire—mankind's most powerful and protean, all-purpose medium to date—not as a single event but as a long process of discovery and development involving many technologies and spanning many thousands of years of prehistory—and it's *not finished yet*. It may never be finished. That helps put the fledgling field of telecommunications in some perspective.

For our ancestors to learn the ways of fire, they must have watched it for millennia. These protohumans slowly overcame their deep and natural desire to flee and began to turn back to observe in wonder. They brought careful witness of lightning strikes, forest fires, and volcanic eruptions—and began to tell the tales of the properties and possible uses of fire, all as prerequisite to working with it: Fire from a log still smoking after a range fire has passed is comforting and warming. Game that has been killed by fire and cooked by it is tasty and easier to eat. Wildfire can be fanned and driven by the wind, and it flames up when it reaches certain materials like dry leaves, grasses, and underbrush. All this, and much more, we would have to have learned before we could successfully tend, use, and keep fire. And then only, having learned how to keep and nurture fire, could prehumans have begun to develop the means to kindle it from scratch with certain sticks or flints and tinder.

We can only imagine how each discovery and technical innovation spread from people to people over thousands of years by conquest, trade, observation, and marriage. It's even possible fire was "discovered" and lost many times over in prehistory. The time scale for watching fire, beginning to use it, and then developing the tools to make it involves hundreds of thousands of years: While paleontologists have found evidence of fire use 500,000 years ago, the first evidence of an ability to make fire is a mere 13,000 years old.

We are still learning new uses of fire: in this century alone, the internal combustion engine for traveling on earth, the jet for air, and the rocket for journeying out into the universe. We've developed the microwave, the laser, and the fire of all fires—nuclear fission. At the same time that we seek ever more intimate uses and control of the medium, we also develop more macro uses of fire, and each new use suggests another, deeper, more powerful possibility for fire. So it is likely to be with telecommunications. The final chapter of this book looks at scenarios of what the long term may hold, where the vector is pointed.

In some important ways after a mere century or two of telecommunications, we are still at the watching stage—although unlike fire, this medium carries the news of itself everywhere almost instantaneously. As with those who first began to watch fire, it may help us to understand some of the technologies for "carrying the fire" that are now becoming available, or will be soon.

Watching Wire

Your phone (or modem) is part of an electric circuit, with a power source at the nearest phone office or station. When you lift your handset from the cradle it closes the circuit, which signals the station to send you the dial tone—the sound that says you are now wired in to the local-regional-national-global network and can place a call to any one of some 700 million phone numbers on earth.

Your local central office or switching station is the hub of your system, which routes your call and initiates the ringing. It makes your connection to another phone if your call is local or to an outside trunk line that sends you to the next station up if you're going long distance. The connection process is called switching. A large city usually has many central offices.

A way to visualize the system is as a set of "stars" or hubs—with all central offices connected to and through each other, up through four levels of switching centers, each encompassing larger geographic areas. The really long connections may be converted to microwave radiation and beamed tower to tower across the country, switched to transoceanic cables strung along the sea floor, or converted to radio waves and relayed by telecommunications satellites to other countries, there to be handled by the local system.

In a nutshell that's how the world is now wired together.

Getting Switched

I talk into the telephone these days as hopefully, though uncertainly, as my brother once talked to horses.
　　　　　　Studs Terkel

Over the past century the switchboard where your call gets routed has evolved tremendously. Ernestine, Lily Tomlin's famous operator character, used the earliest method: When you lifted your handset (or, in the old days, cranked up your phone), you initiated a small "ceremony of the phone call." A light came on over your socket on the operator's switchboard. She (seldom he, in the old days) inquired politely what number you wanted, said, "One moment, please," and grabbed a cord of the appropriate length, rang the party to see if the phone was busy or available, stayed on the line just long enough to hear someone answer, and then plugged one end of the cord in your socket and one in your party's socket. (Ernestine tended to listen

Lily Tomlin as Ernestine. For better and worse, Tomlin's Ernestine has been replaced by all-electronic switches like ESS 5 and some of the most complex software on the planet.

in and even join the conversation if she felt so moved.) When the board lights went out, the customers had hung up, and she could remove the cord. Switchboards like this are still in use in some parts of the world and even in a few remote areas of the United States.

The human operator had the effect of personalizing your trip into telephone cyberspace, and many people protested the end of the Ernestine era. With the rapid growth of the phone system, though, Bell managers became alarmed by projections that showed the need for operators would eventually exceed the U.S. population. It was clear early on that some sort of automatic switch would be necessary, and the first began operation in 1892 in La Porte, Indiana. The electromechanical device was composed of selector switches, or "relays," that moved into the correct position with each digit of a number dialed. The old rotary phones were deliberately designed to move slowly enough so the relays would have time to make their connections.

Electronic switches, which began to be used in the 1960s, don't need the delay. With a touchtone phone, of course, you can dial as fast as you can push buttons—or faster, as in the case of a redial button or speed dialing of preprogrammed numbers. After dialing there's a pause of 7 to 15 seconds before the switch is made and the line starts ringing. With new fully digital services, though, even that lag will all but disappear.

Now state-of-the-art switching is all done by powerful computers and mammoth software programs. As Robert Lucky, vice president for applied research at BellCore, writes, "Today's central office switches like the AT&T ESS 5 are among the most complicated systems made by mankind." The ESS 5 is an advanced office switch controlled by a program composed of about *two million* lines of code and costing about a billion dollars to develop. Over half the code is for maintenance and trouble shooting. The ESS 5 is constantly scanning every phone line on its board (hundreds of thousands) and can complete one scan of the whole board in a tenth of a second. If a problem develops, the computer can reload it's whole operating system in less than 10 seconds.

The upside of these all-digital switches is their speed and durability relative to the old electromechanical devices. They also obviously can handle an enormous amount of traffic (190 million calls on a normal weekday). The downside is their complexity, the difficulty of adding new software features that must interact with so much preexisting code, and the possibility that the software could be tampered with or

AT&T's Network Operations Center in Bedminster, New Jersey.

that it may contain inscrutable errors. Author Bruce Sterling put it nicely: "The faults in bad software can be so subtle as to be practically theological."

In his book *The Hacker Crackdown,* Sterling documents how one such fault caused a major crash of AT&T's long distance switching system on January 15, 1990. The author warms to his topic with this description of AT&T's very bad hair day:

> This was a strange, dire, huge event. Sixty thousand people lost their telephone service completely. During the nine long hours of frantic effort that it took to restore service, some 70 million telephone calls went uncompleted. . . . Station after station across America collapsed in a chain reaction, until fully half of AT&T's network had gone haywire and the remaining half was hard put to handle the overflow.

It turned out that a small glitch in a new version of the operating system that controlled the reload cycle caused the chain reaction right up the line. Luckily, AT&T had kept copies of the old operating system, which it reinstalled while the glitch was sleuthed.

The story illustrates the immensity of network activity and how central it has become to the way our culture and commerce function. It also offers a cautionary tale about the potential fragility of a system that relies on something as ephemeral and "theological" as software. The speed and capacity of the new "pipes," like fiber optics, will increasingly challenge even the mightiest of switches.

The Pipes

Fiber Optics

Although Alexander Graham Bell experimented with light transmission, it wasn't until the technology of fiber optics, now about 25 years old, that it became feasible and then preferable.

You may have seen a crude version of optical fiber in a lighting store: sheaves of clear polymer spaghetti with a bright glow at the tip of each strand. A light source at the base of the sheaf sends light up each strand, which mysteriously "contains" the light, even around a slight curve, until it appears again brightly at the tip.

Telecommunications fiber is made by drawing out long, hair-thin strands of ultrapure fused silica, the primary ingredient in sand. Depending on the quality and purity, these fibers can transmit light

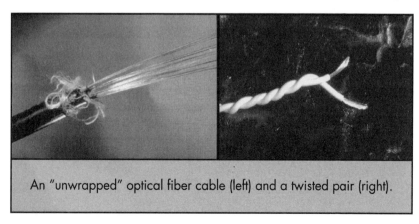

An "unwrapped" optical fiber cable (left) and a twisted pair (right).

some 7 miles or more before it needs to be amplified. An analog phone signal is converted into superfast digital pulses of light transmitted by a laser. At the other end the signal can be reconverted into an analog signal or amplified and sent further.

The individual fibers are bundled into a cable, such as the ones used in AT&T's Northeast Corridor Network, running from Massachusetts to Virginia, which has installed fiber cables containing as many as 50 fiber pairs. Most of the country's main phone traffic corridors are now similarly wired with fiber.

Commercial fiber systems that span the United States transmit at a rate of 1.7 gigabits (billion bits) per second, per fiber—the equivalent of about 25,000 voice channels. But that's just the beginning. The capacity of optical fiber to transmit information has been doubling every year since the 1970s, when the first major fiber optic line was installed. Although materials scientists are steadily improving the "transparency" of the silica, the doubling is not accomplished by changing the installed fiber but by improvements in the electronics: the laser transmitters and the photo-detector receivers.

Recent laboratory experiments have shown fiber has a capacity of more than 50 times that, or 100 gigabits per second—truly a firehose of bits. One million television channels could "fit" in such a fiber. That's much more capacity than we need now, but just wait. Two-way high bandwidth connections into most homes and businesses could use it up. The oft-heralded "video on demand," wherein you select the film you want whenever you want it and *Voila!* it's there (with all the VCR functions like pause, rewind, and fast forward) is, naturally, very bandwidth intensive. And real-time, two-way, interactive multimedia into many homes and businesses could use up a lot of capacity very quickly.

In spite of the capacious advantages of fiber optics, copper telephone wire (often called twisted pair) is still the most common connection into homes and small businesses, and there is a huge installed base. Although twisted pair is considered old-fashioned and low bandwidth, its upper capacity of 6 million bits per second is only just recently beginning to be exploited. Modems could operate 100 to 200 times faster than they do today over the existing copper lines, and these same lines could even transmit VHS-quality TV signals.

But fiber is clearly the "lightwave" of the future. While fiber gets cheaper and faster, copper is getting more expensive. In fact, in China, authorities prefer fiber because some villagers tend to dig up the copper wire to sell on the black market—which can be very hard on the infrastructure.

Fiber has other advantages beyond its lower cost. It's not affected by electromagnetic disturbances. And because of its giga-boggling bandwidth, much of the newly installed base of fiber currently sits idle. It's referred to as dark fiber, awaiting future need and demand. It will come.

Telecommunications Satellites

Since the early 1960s the satellite system has expanded to carry most of international phone traffic. Satellites are also a major and growing relay source of television for developing countries and remote areas of the world.

In 1945, Arthur C. Clarke, the novelist and screenwriter for the classic science fiction film *2001: A Space Odyssey*, first proposed the idea of using a satellite as a relay station for communications. Nearly 20 years later, his vision became reality when the first Telstar "bird" was sent aloft with the capacity to receive, amplify, and relay one television channel or several hundred voice channels. SynCom 2 followed shortly, the first geosynchronous satellite, "parked" in orbit above a particular spot on the earth's surface. The satellite is accelerated to the right velocity and height until it precisely matches the earth's rotation though some 22,000 miles up; it is still an orbiting captive of Earth's gravity. Although it appears stationary from Earth, it's actually moving at a tremendous velocity to "keep up." (Think of the difference in speed between the inside of a merry-go-round and the rim.)

High-altitude communications satellites can "see" a huge swath of Earth and so can send and receive signals to and from many earth stations simultaneously. Because of the potential for crowding of telecom traffic on the frequencies that have been assigned, signals are now beginning to be converted to bits. The digital capacities are five times greater than the old analog transmissions.

A contemporary satellite like AT&T's Comstar can relay as many as 18,000 two-way telephone conversations or 24 TV channels. Some of the international telecom satellites put up by Intelsat, a consortium of the United States and over 100 other member nations, have an even larger capacity—up to 100,000 phone channels.

Car interior with navigation system on the dashboard. The driver's route is shown onscreen.

Satellites are also now used routinely for communications to and from ships and planes across the globe. The NAVSTAR Global Positioning System and its network of 24 satellites allow boats, airplanes, wilderness adventurers, and anyone else with the right equipment (now available as an inexpensive hand-held unit) to check their exact location on Earth to within a few meters. The technology has already changed the art of map making. And prototypes of an intriguing system for finding your way around town have already debuted in some rental cars in Los Angeles and in some luxury automobiles. A computer in your vehicle, equipped with map software working in conjunction with a GPS unit, shows you an icon of your car moving along the map toward the destination address you've entered. The software calculates and highlights the ideal route. Make a wrong turn and you can see it instantly, as well as an update on the *new* best way to get there. To help you keep your eyes on the road, some systems feature a synthesized voice that alerts you with messages like "Turn ahead" or "You are off route." This system may eventually come as optional or even standard equipment for all cars. (One drawback: You'll never again be able to excuse your tardiness with the line: "I got lost.")

No Strings Attached (or Plugging into the Air)

Spawning jokes about no phoning sections in restaurants, the cellular phone (also called the mobile or radio phone) is now a ubiquitous fixture of U.S. and global technoculture. The mobile phone system keeps you connected while you're on the go by dividing up service areas into "cells," each served by a base station that sends and receives your call as a radio signal and routes it into the regular phone system. As you move out of range of one station, the system automatically switches you to the system cell area you are entering.

Following on the heels of cellular phones are the personal digital assistants—small portable devices with touch screens and graphical user interfaces that add fax, text, and graphics (and eventually video) to the wireless voice phone. The era of Dick Tracy's wrist phone may also be right around the corner, and AT&T has already unveiled a prototype.

In 1983, industry analysts couldn't see more than about 1 million people using cellular phones by the year 2000. As it turned out, some 50 million were already in use in 1995, and the number has been growing by 50 percent a year in North America and by 200 percent in parts of South America. Analysts now predict that three-fourths of all U.S. households and a half billion people worldwide will be using wireless by the year 2000. Because of this extraordinary growth, cellular services, which use radio frequencies and were designed as analog systems, are jumping on the digital bandwagon for the same reasons as all other forms of telecommunications: increased capacity, the ability to compress data, and improved transmission quality through error correction.

AT&T's wristphone prototype.

Developing Without Wires

For someone living in the developed world and taking the ubiquity of phones for granted, it's astonishing to learn that over half the people on earth have never made a phone call—mainly because they don't live anywhere near a phone or phone line. Some developing countries with little in the way of traditional phone infrastructure are skipping the wire phase and going straight to wireless to offer their citizens regular phone services. The cost of building a wireless network is much lower and can be done quickly. Argentina, for instance, recently built a countrywide wireless system, and some remote areas, which probably could never be reached by wire, achieved phone service in less than six months.

Alphabet Soup

One inescapable feature of our technological age is the proliferation of jargons for each new field. Experts seem to take an almost perverse delight in creating impenetrable means of communicating with one another. And telecommunications surely ranks at the top of the pile for sheer mass and density of jargon. Anyone who dares to read prognostications about the field is quickly engulfed in a blizzard of acronyms. The following is a small taste of e-mail dialog from a discussion in an on-line telecommunications forum:

> ATM will, for some time, support legacy protocols through tunneling, encapsulation and emulation environments, i.e., ATM will be used to transport Ethernet, Token Ring, SNA, TCP/IP and IETF-contrived data bundles, but over time, perhaps five years from now, native ATM will have already displaced many of these embedded technologies.

And that's actually a lucid statement compared to some telecommese prose. Although most of the jargon can be left to the propeller heads, there *are* a few acronyms and terms that you, as a budding telecom "watcher," may need to deal with.

Bit

Bit is short for "binary digit." It's the smallest unit of information a computer processes. It can represent on or off, high or low, yes or no, 1 or 0. It's the digit in the word "digital."

Bandwidth

Bandwidth is the range of electrical frequencies or the amount of digital information a device or line can transmit in a given amount of time. A

VHS-quality video channel, for instance, requires very high bandwidth—the equivalent of about 1,200 voice channels. Sometimes bandwidth is expressed as bits per second.

Compression

Data compression is exactly what it sounds like, squeezing down the spaces or repetitive sections in a digital transmission. A compression algorithm is a software formula used to compress data. Example: "Computer: Every time you encounter the word 'antidisestablishmentarianism' substitute 'A+'."

ISDN

ISDN stands for Integrated Services Digital Network and was invented ages ago (in 1978) at Bell Labs. ISDN was designed to be the digital replacement for conventional analog telephone service, using the same infrastructure of standard (copper) phone lines, but offering new capabilities like simultaneous transmission of voice and data. Because it lay on the shelf so long, telecom wags claimed ISDN stood for "It Still Does Nothing." Now, though, the technology is finally beginning to come into its own as a transitional service while the blazing fast, high-bandwidth infrastructure of the twenty-first century gets worked out and put in place. Chances are if you call your regional phone company, it will be able to quote you a price on an ISDN hookup.

A growing number of companies use ISDN to enable telecommuters—employees working part- or full-time at home from computers—to tap into the company's local area network (LAN), communicate with colleagues at the office, and use all of the LAN's other broadband features. Lawyers can send and receive large briefs at home and utilize the firm's legal on-line information services. They can also speak with other attorneys or paralegals while viewing the same file and making changes in it. Hospitals transmit digitized x-rays and other diagnostics to service providers and specialists. A composer who creates music and sound effects for a nightly news show can send a CD-quality transmission of his work each day in time for the edit at the studio. The rise of the Internet's World Wide Web has also spurred the growth of ISDN because it allows the transmission of data at 128,000 bits per second, more than four times the speed of the fastest modem. That's a big difference if you have a serious need to down- or upload high-resolution graphics, sound, or video.

Other features of ISDN:

- High-fidelity voice. The low-bandwidth, low-quality telephone voice we're used to is but a shadow of what it could be—improving the intimacy and clarity of phone conversations.
- Call waiting, caller ID (you see the number of the person calling), and e-mail.
- Collaborative computing: You can simultaneously talk and work on a single document with one or more colleagues.
- Desktop videoconferencing—the long-heralded video phone may finally be within reach. Talk to and see one or more colleagues on your computer screen while a small camera mounted on your screen sends your image over the wires.

When reading about ISDN you may come across another acronym: ADSL, for Asynchronous Digital Subscriber Line, which is a different technology than ISDN but enables similar capabilities and services. Because it potentially offers even greater capacity and has other advantages, ADSL may supplant ISDN as the technology of choice for offering advanced services over the existing copper wire infrastructure.

ATM

Now you have not only your friendly automatic teller machine but also the more daunting asynchronous transfer mode. This opaque phrase is a new digital transmission and switching technology that most of the world's manufacturers of computer and telephone equipment have agreed to use.

ATM allows very high-speed, efficient, and flexible use of the network capacity. Instead of holding a line or signal path open during a phone call or transmission, and tying up space whether or not anything is being transmitted, ATM avoids this by breaking the transmission into discrete "chunks" of data.

The technology converts the signal, whether voice, text, graphics or video, into 53-byte (= 424 bits) packets or cells, each of which has an address and information identifying what's in the cell and how and when it needs to get where it's going. These cells can be flung through the same signal path, or line, and combined with cells from other transmissions to make maximum use of an available channel. And cells from the same transmission can be routed differently, as long as they end up where they're going in time to be reconstituted into the original message or data.

Ultimately, the efficiencies of ATM (or a packet-switching technology like it) will enable on-demand use of the pipes for any kind of telecom use we can imagine. And thousands of uses *are* being imagined.

Big Cable

Although the telephone network is the most ubiquitous, completely intertwined telecom infrastructure, it is not the only player. Cable television companies control an immense and growing installed base of very-high-bandwidth cable. Some 63 million homes have a cable hookup, and consumers spend $25 billion a year on cable services. Although most of the cable system was not installed with interactivity in mind, a few companies have developed the technology to provide high-speed Internet hookups to subscribers. This requires that the user purchase some fairly expensive equipment (such as a cable modem). A few pioneering satellite TV companies are also offering similar services.

The deregulation of the industry through the Telecommunications Act of 1996 opens up some quite interesting possibilities for two-way interconnectivity exploiting the broad bandwidth of cable and the sophisticated switching and worldwide reach of the phone system.

One service that has these companies' mouths watering is video on demand (VOD), which is what it sounds like, video when you want it. At first, this will probably be a small selection of current films. Eventually, it could expand to include whole libraries of films, any of which can be summoned and operated with the VCR-like controls we've grown used to, like pause, fast forward, and rewind. When this happens the video stores will presumably be out of business. As Nicholas Negroponte points out in his book *Being Digital,* $3 billion per year of the $12 billion video rental business is in late charges.

Having a movie served up in an all-digital format seems at first blush to be just a more efficient method of piping films around. But a digital film is in some important ways a different medium from a film on tape. The fact that you can jump instantly to any point in the film gives it

> ## "The Negroponte Switch"
>
> Nicholas Negroponte, an information guru with a very high profile, is a co-founder of MIT's famed and well-funded Media Lab and a strong proponent of telecommunications consumers having the maximum possible choice in media programs and services. Long an advocate of all-digital communications, he suggests that if subscribers have a digital conversion set-top box on their TVs with a storage capacity and high-bandwidth connection, the film of one's choosing could be downloaded in a few minutes, or less. And why stop at movies? he asks. He envisions all television programming being accessible on demand. You view the news and any other programming when you want them as opposed to when they are scheduled.
>
> Negroponte has also called for reversing the ways we use our telecommunications infrastructure. It would make much more sense, he argues, for broadband communications that are directed to stationary devices like home televisions to be sent via wire and cable. This would free up the airways spectrum for communications, like telephone, that need to reach individuals wherever they are. This proposed change has been dubbed the "Negroponte switch."

some of the utility of a book. Films may be packaged with invisible headers to each scene, which can be viewed and used as a screen menu. You may be able to add your own indexing with a click of your remote as you watch the film. In general, an all-digital format should make it vastly easier to use film as reference material, as well as for passive linear entertainment. A service may evolve that allows you to file your favorite scenes for research or review. Another could provide viewers with personally specified film collages from the whimsical (like a half hour of back-to-back clips of famous film monsters acting ferocious) to the academic (such as four different actors performing their film versions of a Hamlet soliloquy for an English or drama class).

Even with popular new digital features, though, video on demand is

A French woman using a standard home Minitel set.

still only an example of what Paul Saffo, a director of the Institute for the Future, calls "paving the cowpaths"—merely using a new technology to offer an improved version of something that is already widespread.

More interesting to consumers than news of Company X absorbing Company Y in hopes of capturing the VOD market are the innovative, perhaps still unimagined, kinds of services and capabilities that will evolve out of the new infrastructures. One possible outcome of deregulation is that our various electronics communications devices will begin to share each others' capabilities. Cable TV will be accessible from your computer, which can also serve as a phone, fax, and answering machine. You'll be able to surf the World Wide Web through your TV sitting on your couch operating a remote (World Wide Web prime time?). Certain Internet services may become accessible via the telephone. The French have enjoyed over a decade of networked services via the Minitel system—widespread in homes and businesses. In addition to such obvious services as on-line white and yellow pages information (long overdue in the United States), thousands of other services are available to French phone customers. Although the interface is an antique now, the Minitel's success offers a tiny preview of the new kinds of services, programming, multiuser games, and utilities that are sure to spring up as a result of deregulation.

CHAPTER FOUR

The Net, the Web, and the Highway

In the many thousands of hours that Mike, a college freshman in Kansas, has been logged on to his favorite MUD, he has created an apartment with rooms, furniture, books, desk, and even a small computer. Its interior is exquisitely detailed, even though it exists only in textual description. A hearth, an easy chair, and a mahogany desk warm his cyberspace. "It's where I live," Mike says. "More than I do in my dingy dorm room. There's no place like home."

Sherry Turkle, *Life on the Screen*

he big news, of course, about the international phone network is its exploding use as the infrastructure for the Internet, the computer network that links all other networks—academic, business, government, and, increasingly, people working and playing from their homes. Part of the renowned robustness of the Internet, its ability to change and adapt, is that it was designed to survive the worst possible contingency: a nuclear war. If one line or host computer was destroyed, other computers could route around the damage and keep the network functional. The Internet evolved from the ARPAnet—founded in 1969 by and for the Pentagon's Advanced Research Projects Agency. It was conceived as a national computer network to link scientists and universities doing defense research, computer science, and other government-funded projects. Over the network they could exchange information and data, collaborate over a distance, and even co-author research papers. The early users, often computer scientists or engineers, tended to be quite familiar with computers and the fairly arcane operating and command protocols of UNIX—the language that got you around in the mainframe host computers and minicomputers like Digital Equipment Corporation's famed VAX line. Soon other academicians and their students were using the ARPAnet regularly for work and even to socialize in the long, late hours. In the 1980s, recognizing that ARPAnet had become much more than a communications vehicle for government researchers, the National Science Foundation sponsored the creation of the Internet as the network of all networks. The well-being and growth of the Net is watched over and administered by the Internet Society and the Internet Engineering Task Force (IETF)—a largely volunteer group of netizens who establish standards and protocols for the Net, see that it keeps running, and plan for the future. Here's a brief look at some of the most popular utilities that have been created for the Internet (the next chapter zooms in closer on some of these):

- *Bulletin Board Systems (BBSs).* These are centered on a topic of interest. You can log on and post and retrieve messages from an electronic bulletin board. Some BBSs have moderators and some are unmoderated. The system operator (sysop) watches over the system and may actually be the one who contributes space on his or her computer for the BBS. The sysop may also serve as moderator, clearing old messages and enforcing rules of discourse. Currently, there are some 60,000 BBSs worldwide. Many of these are accessible independently of the Internet. If you have a computer and modem and know the protocols, you can connect directly to a particular BBS, even without an Internet account.
- *Usenet.* In 1979, several graduate students at the University of North Carolina invented the Usenet Newsgroup system as a means to distribute news and information about the UNIX operating system. Today, there are 14,000 Usenet newsgroups, generating millions of messages a year on every imaginable human interest, including religion, politics, sex, cooking, sports, and warfare. As a subscriber to one of these, you (or your workgroup's file server) can receive all the messages and posted news about a particular ongoing topic of interest and post your own replies or bulletins. Unlike a BBS, the news flows to you automatically, and there is no central computer administering an individual topic. In his book *The Virtual Community,* Howard Rheingold characterizes Usenet as an "anarchic, unkillable, censorship-resistant, aggressively noncommercial, voraciously growing conversation among millions of people in dozens of countries."
- *File Transfer Protocol (FTP).* This program was developed to allow users to automatically download text or software from Usenet

groups, BBSs, or other archives on the Internet. Originally, it took a bit of a learning curve and special software, but many of the online services and the World Wide Web have reduced file transfer to a simple menu or icon selection.
- *E-mail (electronic mail).* This is the most widely used of Internet utilities. Futurist and Net watcher George Gilder estimates that in 1995 e-mail traffic first passed postal service traffic by 95 billion to 85 billion units. With an internet e-mail account you can correspond with any of the millions of other netizens. The advantages over traditional mail (now dubbed "snail mail") are several. Convenience—you type your message in the e-mail utility window and, when it's finished, send it immediately with one command or a click on a Send icon. Speed—usually your correspondent receives your message within hours, if not sooner—sometimes within a few minutes. Cost—there is no cost beyond that of your Internet account and monthly phone service. No matter where in the world you are e-mailing, each transmission counts only as a local call. There are also several disadvantages: E-mail currently has a generic look, plain text, and no letterhead. Getting a real letter is still a nicer experience. Although plenty of business is done through e-mail, it has a very informal quality, like passing notes in class, or a cross between a letter and a voice mail message.
- *Talk.* As an Internet subscriber you can send a Talk invitation to any other netizen in the world. If they're logged onto the Net, your invitation appears on their screen, and if they're willing, you can begin a typographic "conversation." Each time you hit Return, your words transmit to your talk partner, and the ongoing conversation scrolls up the screen.
- *Internet Relay Chat (IRC).* IRC expands the Talk feature to real-time chatfest with several or many participants all chiming in. Each new comment, question, or response appears at the bottom of all participants' screens and scrolls upward. The first IRC program was written in 1988 by Jarkko Oikarinen at the University of Oulu in Finland for use there. As with so many other useful Internet utilities, it spread worldwide within a year. IRC can be and is used for academic or business conferencing. Compuserve and other on-line services regularly feature a guest expert in a particular field, or a panel of experts, and any subscriber can go to the conference room,

The Internet as Boomtown

From July 1994 to July 1995, the size of the Net, as measured by its host computers (or file servers that store, dispense, and route data), doubled from 3.2 to 6.6 million. That doubling is projected to continue until by the year 2000, 120 million hosts will exist. (World Wide Web servers skyrocketed from 130 in June 1993 to more than 40,000 at the end of 1995.) More than half the hosts exist in North America, with the others spread through more than 100 countries worldwide. Growth is now greater outside the United States than inside, a sign of the Net's increasingly international character.

Some 60 million people were using the Net at the end of 1995, and that number will grow to 200 million at decade's end.

A demographic snapshot of Web users: 77 percent are male and 23 percent female, but gender use is balancing out rapidly. Median income of users is $40,000, but 28 percent make less than $20,000. Total sales over the Web were $118 million from September 1994 to August 1995.

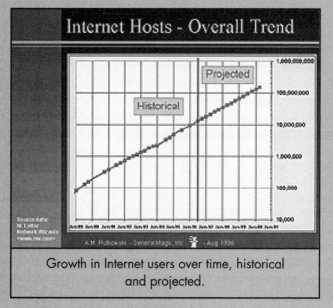

Growth in Internet users over time, historical and projected.

ask questions, and chat IRC style. But what IRC is most often used for is a sort of verbal "playground" where people go to hang out to engage in any number of ad hoc conversations and repartee. Howard Rheingold likens it to "the corner pub, the café, the common room—the 'great good place' of the Net."

- *Multi-User Dungeons—or Domains (MUDs); Multi-User Adventures (MUAs); MUD Object Orienteds (MOOs).* Some of the most elaborate and strange kinds of communities on the Net are centered around various incarnations of these groups. Communications scholars have made these phenomena the subject of dissertations and whole books. MUDs arose as a sort of networked version of the gaming phenomenon generally called Dungeons and Dragons. You exist in these domains as a persona with a nickname. Players amass experience and power over time, going on quests, picking

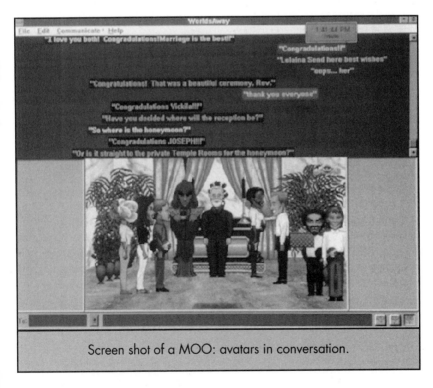

Screen shot of a MOO: avatars in conversation.

up objects, trading with or battling other players and armies, and using the MUD's computer language to build their own objects, weapons, charms, and spells. The MUD environments and action are all communicated via text and descriptive passages. While these games are not for everyone, especially because casual "newbies" have a tendency to be killed, some players spend as much as 70 hours a week on this activity. As computers have developed more graphical capability, a visual variant on MUDs, called MOOs, has been on the rise, in which players can select or create cartoon-like graphical representations of their characters, dubbed avatars. In these games you may be able turn your avatar's head and move its eyes to regard another avatar; and when you "say" something the message appears in a balloon above your avatar's head.
- *The World Wide Web.* For most newcomers since 1993, the World Wide Web is all they'll ever need or want to know about the Internet. This is especially true since the Web can and often does incorporate all previous useful or popular features of the Internet, such as FTP, Talk, Chat, and access to MUDs, MOOs, and the Usenet, and makes them easily available to anyone who can point, click, and enter text on a computer. Early versions of speech and video telephone capabilities have been added to the Web as companies race to add functions and features to it. More on the Web further on.

What's Up in the Road, Ahead?

So is the information superhighway simply a synonym for the Internet? It all depends on who's talking about it.

Bill Gates thinks it doesn't exist yet—that the Internet is not the highway. He thinks it will be built by a company or consortium that believes it can make back over time, through fees and services, the $120 billion he estimates it would cost to connect American homes and businesses with a high-bandwidth infrastructure like optical fiber.

George Gilder thinks that it already exists in a nascent form as the Internet, which will be improved, upgraded, and expanded incrementally over time as a consequence of market demand. For many years before the Telecommunications Act of 1996, he was a strong proponent of deregulation of the communications industry to allow cable TV and

Microsoft chairman Bill Gates.

phone companies to explore each others' territories either in alliance or solo. Combining the extant and growing broadband cable infrastructure with the interconnectivity and global reach of the phone system can offer us all or many of the interactive multimedia features envisioned for the superhighway—what Gilder calls a "broadband Internet."

The Computer Science and Telecommunications Board of the National Research Council calls the highway the "National Information Infrastructure" (NII).

A national information infrastructure can provide a seamless web of interconnected, interoperable information networks, computers, databases, and consumer electronics that will eventually link homes, workplaces and public institutions together. It can embrace virtually all modes of information generation, transport and use.

The Board calls for maximum openness in its construction: open to users, open to service providers, open to network providers, and open to change. It also distinguishes between the NII and the Internet.

Virtually none of the info gurus seem to like the metaphor of the "information superhighway." In his book *The Road Ahead*, Gates says that a geographical term like "highway" doesn't suit. He says, "In fact, one of the most remarkable aspects of this new communications technology is that it will eliminate distance. It won't matter if someone you're contacting is in the next room or on another continent, because this highly mediated network will be unconstrained by miles and kilometers."

Gates, not surprisingly, proposes a quite different metaphor—"the ultimate market":

> Markets from trading floors to malls are fundamental to human society, and I believe this new one will eventually be the world's central department store. It will be where we social animals will sell, trade, invest, haggle, pick stuff up, argue, meet new people, and hang out.

George Gilder

George Gilder is as upbeat as anyone on the power and potential of the Net and the Web. Gilder, currently a fellow of the Seattle-based Discovery Institute, feels that Moore's law, which has accurately predicted the doubling of computer power every 18 months for the past 30 years, is actually too conservative. Because of the synergy of having ever greater processing power, Gilder feels that the effect is more like a quadrupling of performance. He also cites Metcalfe's law (named after the inventor of Ethernet), which states that when you interconnect computers in a network, be it a local area network (LAN) or the Internet, each computer increases its performance exponentially, again as a function of the synergy of connection and the number of networked computers.

George Gilder.

Gilder is skeptical of the recent mergers/conglomerations of content, conduit, and service provider companies as a false and last hope for monopoly profits. He sees the Net and the Web as a powerful, evolving global system well beyond the control of any individual company or conglomerate. Instead, he envisions thousands

---

Such a megamarket is, of course, also a place where the high stakes poker players of commerce will make fortunes—or increase them. It's interesting that Gates locates this "market" in the future, whereas a growing number of people already do all the things he describes, and more, on the Internet as it currently exists.

Nicholas Negroponte doesn't seem to care what it's called as long as

of companies leapfrogging each other to increase the bandwidth of the Internet and make it more functional. In a recent interview in *Internet World* magazine, Gilder says, "My prophecy for the last five years has been that computer networks, epitomized by the Internet, would usurp both television and telephony, and that is on its way to happening."

Gilder also predicts that the Web will effect what he calls "a hollowing out of the PC." With a relatively inexpensive computer—some have dubbed them "Net-Tops" or "Web PC's" because their primary function is to access the Internet and World Wide Web—anyone will have access to software, information, and resources that would presently cost thousands. Users may not even own software anymore, but rather rent it for a particular task. It will even be possible to store your work and files on a server somewhere on the Net, for a small fee.

One of the keys to this new model of computing is being able to automatically meter transactions on the Net, making it possible to charge users mere pennies to read an article or download an item. Some transactions might be valued at only fractions of a cent, what Ted Nelson calls "nanobucks." Says Gilder: "As that kind of technology is perfected and incorporated into the Net, it will be a much more prosperous place. At the same time, it won't impose impossible burdens on the information-poor. Instead, it will make it possible to extend services to the poor that can't currently be extended because of the granularity of the charging mechanisms required." Companies won't necessarily be offering these services purely out of the goodness of their corporate hearts. As Gilder puts it, "The biggest untapped market is always the poorest."

it's digital and available to everyone. In the conclusion of his book *Being Digital*, he writes:

> The information superhighway may be mostly hype today, but it is an understatement about tomorrow. It will exist beyond people's wildest predictions. As children appropriate a global information resource, and as they discover

MIT's Nicholas Negroponte

that only adults need learner's permits, we are bound to find new hope and dignity in places where very little existed before.

Highway Ends in ¼ Mile

The only safe prediction is that none of the pundits', nor anyone else's, scenarios for the superhighway will get it 100 percent right. Something of all of them will survive. The infrastructure will co-evolve to serve the capabilities of the new and completely unpredictable "cowpaths" of the Internet.

In the process, hybrids of new media will arise that are surprising, even astonishing. The World Wide Web caught nearly everyone by surprise. It is now a new megamedium in its own right, created of other media—many of them quite recent. The Web in turn is becoming a vast spawning ground and test bed for a splurge of other new media. As the capacity of the pipes enlarges dramatically, the nature of what can be and is being sent back and forth through them is evolving in predictable and also extraordinary ways.

CHAPTER FIVE

What's New About New Media?

Software is cool.
            Bill Gates

We become what we behold.... We shape our tools, and thereafter our tools shape us.
            Marshall McLuhan, Understanding Media

Mediation, mediation, mediation. I want an end to mediation.
            Sven Birkerts, The Gutenberg Elegies: The Fate of Reading in an Electronic Age

Barrier Bridge

A medium is a paradoxical entity. It is something by which or through which we obtain a certain experience, and it also stands "between" us and that experience. So, it's both the means to something and a barrier to it.

The phone allows me to talk with my brother across the country. It can also seem to stand between us and the experience of conversing face-to-face. The gestures, the reactions, the smiling are missing. Anyone talking on the phone with a lover has occasionally felt acute frustration at not being "there." This feeling of mediation or separation creates a tension or dissatisfaction. And this dissatisfaction is part of what drives scientists and engineers to attempt to heighten the experience by improving or innovating the medium. Better sound quality or a video image will increase the sense of presence and "immediacy." But there will still be a tension between the mediated experience and the experience the mediation attempts to recreate.

Film and TV bring people onto screens in the theater or into our homes—but it's a one-way communication. It doesn't bring us into their lives. Our experience with these media is also that of a relatively passive consumer, and we watch the movie according to the schedule of the screening or broadcast. Interactive multimedia increase our level of choice. Even a videotape gives us new options: We can play it at our leisure, watch parts of it again, or fast forward through boring portions. But even the most advanced interactive multimedia are about "what it will be when" as much as what it is now: better sound, higher-resolution graphics, more immediacy.

Media Sophisticates

Since the Industrial Revolution, and at an accelerating pace in the twentieth century, we have become extraordinarily sophisticated and adept at using hundreds of forms of media.

For instance, take the ancient medium of fire: In addition to the venerable campfire, these days we harness dozens of other forms of fire. We now operate a stove or oven to cook our meat. We even have a sly, fireless version of fire in the microwave. We banish the night with the slow, bright burn of filaments or gases in a wide variety of lights—from house lights to flashlights to the tiny fires of diodes. When it's cold, we turn up the thermostat, and soon it's as warm as a spring day. With the twist of a key we ignite the well-hidden fires of the internal combustion engine in our automobile and use fire to travel where we will. We light a candle or strike a match and recreate the archetypal flame fire in hearth or barbecue pit.

Our sophistication with communications media has also grown immensely. In addition to our reading abilities (a book, a newspaper, a comic strip, all require somewhat different reading skills), we have, in the age of cinema, built up a largely unconscious fluency in the grammars and conventions of recorded sound, film, and video.

- We know that a radio is not the voices of ghosts or ancestors from the spirit world—a common first impression of aboriginal peoples.
- We know that the TV or movie screen is not a window, and we no longer bother to look behind it to see where the people are, though we might have done so when we were very young.
- We know we're watching images of people and events captured

previously, and that what we're seeing is not happening in "real time" unless it's a "live" broadcast.
- We generally know that a feature film, no matter how realistic, is a fiction and did not really happen, but if it's very good we suspend our disbelief and experience it almost as though it *is* happening.
- On the other hand, we know that a documentary is "real" but that it probably has a point of view—a selective reality that may even be quite misleading about what *really* happened.

If there's a shot of an automobile moving on the freeway and we suddenly find ourselves inside a car eavesdropping on two people talking and driving, they are very likely inside the car we saw in the prior shot. If the scene of the car interior fades to black and then fades up on the car parking, we know that some time has passed and the people reached their destination. Film makers can use cues that indicate the passage of time from a few minutes to a day, weeks, months, or years. We aren't confused by changes in perspective, like the screen jumping from a close shot of someone's face to a wide shot of that person in a crowd, or by the screen jumping from one face to another during a conversation between two or more people.

There are dozens of other musical, audio, and visual cues we've learned that indicate various levels of danger, humor, affection, joy, and other emotions.

And our cinematic sophistication has been growing and changing. In the old days if two actors got into a fight we saw them duking it out in a single wide shot as though on a stage. Now, no director is content with that. We are dragged right into the middle of the fight, with a fast series of jump cuts, close ups, special effects, and sounds. These new techniques are part of the reason for the resurgence of concern about violence in films. The new violence is *more* violent, more visceral. In fact, you may leave the theater feeling that *you* were the one who was mugged, not just the actor.

The speed of cuts is still increasing. Viewers of MTV and modern advertising have learned to process several minimally related or random shots per second—an editing speed that 40 years ago would have sent viewers jumping out of their skins or to bed with a headache, if they could even have watched it at all.

Computer-mediated communication is much younger than the cinema. But as new conventions are developed and we become fluent in them, we will, over time, also grow more sophisticated and demanding of the interface.

What's New About New Media?

The ideas of Marshall McLuhan, who believed that the biggest story of the century was the hidden powers of media, have had a recent resurgence. This is perhaps because in the digital age the power and proliferation of media are becoming more apparent to the rest of us. For some, the computer and computer-mediated communications have become the central feature of their lives.

"New media" is sometimes used as a synonym for interactive digital multimedia. I think of new media as any and all of the unpredictable hybrid tools, capabilities, arts, and entertainments we're fashioning daily out of the bits and bytes of this new digital universe. The recent ability to digitize all communications media has made these media much more plastic and dynamic—in the sense of malleable and changeable. The realm of the digital is very like the world of dreams, where, for example, a color can also be a flavor and a sound.

Contemporary video editing and music composition are especially illustrative of the changes wrought by the personal computer and its GUI (graphical user interface). A composer today who works with computers has an incredible arsenal at her fingertips. And she no longer has to wait until an orchestra or a band is assembled and rehearsed before she can hear her composition or perform it for others. The so-called orchestra in a box gives her access to all the "voices" of traditional instruments, as well as an infinite range of other sounds. She can play her composition on a keyboard hooked up to the computer and have it automatically transcribed on the screen as musical notation. She can assign an instrument, such as an oboe, to a group of notes or to a whole line and then switch it to the sound of a cello, or any other sound that pleases her. As she works she can select any portion of the notation and instantly hear it played back. If she wants to change the pitch or length of a note in a melody, she changes it on the screen, and when she plays it back the melody is altered as written.

In the digital realm, the sounds of two or more instruments can be blended—a saxophone and a bell become a "bellophone" (or a saxobell). And the composer can view the music and tones in a variety of ways: as notes on a staff, bars, lines, numbers, colors, or even wave forms—all of which can be tinkered with to change the music in minute or dramatic ways.

The art and mechanics of editing film and video have been similarly changed. Formerly, the main option for putting together a videotape was by assembly editing, which involved transferring scenes and shots in order, one by one, to an edit master tape. If you decided later that you didn't want a particular shot in the middle of your master, you paid the

Computers and powerful software are completely retooling the arts. Shown here are some onscreen "palettes" available to composers.

A videotape editing suite.

penalty of reassembling the whole tape. The new world of digital (also called nonlinear) editing has removed that particular hassle, and many others. The fact that the shots can be cataloged and stored as digital information on hard discs means that they are instantly available to work with and manipulate. The software gives video editors the same freedom to make changes in their compositions that the word processor gave writers.

Editors can see a visual outline of the whole piece or view it as a series of frames, like a film strip. They can quickly play back any sequence or watch the whole thing. If they want to remove a frame or ten or a whole shot, they mark the "in" and "out" points and delete, leaving no blank stretch in the playback. Inserting shots is as easy and as penalty free as removing them. The transitional effects between shots and scenes, such as fades and dissolves, are easy to accomplish—and hundreds of other effects are quickly available in the nonlinear studio.

Composers' tools and nonlinear video editing typify the many software inventions used to create interactive multimedia, also known as new media.

Disc Magic

Until recently, when people thought of new media, CD-ROM came to mind. It is certainly the first successful commercial vehicle to converge all traditional media: the spoken word, the hand-lettered word, script and calligraphy, typography and typographic effects, the designed page or screen; all forms of instrumental and electronic music, singing, natural sounds, and sound effects; all graphic media, photographs, maps, signs, drawings, paintings, film, video, 2-D and 3-D animation, cartoons, and special effects.

And all of these media can be used naturally "as they are," or they can be digitally enhanced or altered.

For the writing and production team, that's a daunting and astonishing palette to work with, and yet all of those media could be used in a single film or video—and occasionally they *have* been. A big part of what makes it new is the interactivity, the ability to explore at your own whim and pace—not merely as a passive consumer of education, information, or entertainment. In an important sense, you become an active co-creator, or collaborator with the authors, of your experience.

In spite of the extraordinary capabilities of CD-ROM, this product is seldom fully exploited, and much, if not most, of it is plodding and mediocre—or worse. The really bad stuff is known as "shovelware" in the trade, because someone has done little more than "shoveled" some information or the contents of a book onto disc. Some CD-ROMs, though, are quite wonderful. The best-selling mystery-adventure game *Myst* was one of the first products to demonstrate the promise of the new medium. There are other standouts, particularly in the field of education and "edutainment" for kids. Broderbund's Living Books series is one example. The medium can also add value to the traditional reference work—even above the new ease and convenience of searching for what you want. CD-ROM encyclopedias have added animations and historic video clips to the text, graphics, and photos of the traditional encyclopedia.

How Many Bytes Can Dance on the Head of a Disc?

Currently, a single CD-ROM can contain

- 650 million characters

- or the capacity of 1,500 high-density floppies
- or 500,000 pages of text
- or 10,000 photos
- or many hours of sound
- or 1½ hours of video
- or various combinations of all of the above.

The cost of reproduction of a single disc, if you order in some reasonable quantity, is about a dollar.

Video quality is currently the weakest link, with small window sizes, poor resolution, and the jerky motion of compressed Quicktime video. For this and other reasons, pundits have long been predicting an early demise for the CD-ROM, and nearly everyone agrees that it is a transitional medium to be superseded eventually by some sort of high-bandwidth media delivery system, perhaps via an advanced World Wide Web.

But recently new life has been breathed into the technology that should keep the CD-ROM more than a few steps ahead of anything available on-line. In 1995, the major consumer electronics manufacturers were able to agree on a single format for a new disc technology targeted to be the successor of the videocassette and CD-ROM, and probably the audio CD as well. The new, improved model, called DVD (for digital video/versatile disk), still looks like a CD but is capable of storing a feature-length movie of high picture and sound quality—or seven times the information content of a current CD-ROM: 4.7 gigabytes (4.7 billion characters). A later version is in the works that can contain up to 18.8 gigs or 540 minutes of quality video! That should give CD-ROM developers room to play around in. Of course, consumers will have to spring for new players and disc drives to the tune of $300–$500 per machine.

The Golden Age of interactive multimedia is still ahead. Right now, producing a worthy multimedia product is, in the words of developer Tyler Peppel, "something of a black art." It requires thousands of dollars of equipment, specialized authoring programs, and other software. The process demands the sustained attentions of skilled programmers, video producers, writers, artists, graphic designers, composers, and, of course, the highly stressed and overcaffeinated producer/project manager, one of whose pleasant duties it is to inform the publisher of pending ship-date slippage.

At this point, very few people have all the skills and talents to make a good CD-ROM as a solo venture, although some of the most successful products, such as *Myst* (created by two brothers), come from just such folks. In the future, multimedia authoring software will become far simpler to use. And a generation raised on multimedia, and schooled in its creation, is likely to produce individuals who are multimedia virtuosi—and will compose fluently and naturally in it. In our lifetimes, we may experience some of the early works of a "Shakespeare/Mozart/Picasso/Houdini" of multimedia.

A New Media Maker

In fact, we may not have to wait long for virtuosic multimedia. The generation that came of age with the birth of the personal computer has produced many young people investing their energies and talents in developing state-of-the art CD-ROMs.

Take Luyen Chou, who was raised in New York City, his father a composer and professor at Columbia University. Although he majored in philosophy at Harvard, he maintained his early fascination with computers, programming, and gaming. He returned to New York to teach at the Dalton School, a private school. Because of his strong interests and skills in computing, he was invited to become director of operations of the New Laboratory for Teaching and Learning—an ambitious program to make Dalton a test bed for twenty-first-century education.

Chou helped obtain a large grant that allowed all teachers who wanted to participate to have powerful multimedia workstations in their classrooms networked to the school's file servers. One fascinating piece of software, called Archaeotype, allowed students to simulate an archaeological dig, in which they would unearth artifacts and identify their culture and period and put together a picture of the civilization. Says Chou of Archaeotype:

> When I first saw it, I thought, "This is exactly what I've been teaching about!" Scholarship is exciting because it's detective work. You create a narrative to explain your phenomena. But when you teach history, or anything else, you do exactly the opposite. If you show them primary source material, it's chosen to fit into the story.
>
> Archaeotype gave the kids the objects and asked them questions like Who was here? and What did they do? and then asked *them* to explain the story.

Another innovation of the New Lab was a networked astronomy program that allowed students to participate in sophisticated, real-life astronomy experiments. Among other things, students ran analyses of photographic plates of galaxies taken at the Palomar Observatory. One girl even discovered her own planetary nebula—dubbed "Hillary's Nebula."

As he witnessed the response to the new learning technologies, Chou developed a vision for spinning out some of the group's discoveries and inventions for teachers and students to use elsewhere.

Chou founded a company, Learning Technologies Interactive (LTI). His work at Dalton had attracted the attention of Stephen Brill, president of Court TV, the cable station devoted to transmitting live trials. Chou's fledgling company developed "CaseMaker," a learning CD-ROM companion to the Rodney King case.

Luyen Chou.

LTI received major backing from Time Warner, a major player in new media. As part of the agreement, Chou developed a CD-ROM version of *Bartlett's Familiar Quotations*. Chou and his team added a number of innovative touches, such as expanding the concept of "quotations" to include well-known visual sequences like the Challenger disaster and the Tiannenmen protester stopping a tank. The arrangement with Time Warner also allowed Chou to develop a pet project, an ambitious and original edutainment adventure called *Qin: Tomb of the Middle Kingdom*. One feature of the best CD-ROMs is that they immerse you in an alternate world or new environment where you are a key player—something akin to controlling the events of a dream. *Qin* (pronounced "chin") is, among other things, a total immersion in a culture and historical moment.

The *Qin* CD-ROM employs lush graphics that perhaps even surpass *Myst's*. Players enter and explore the grand tomb of Qin Shi Huangdi (Son of Heaven), the first emperor of China. The actual tomb, located in the province of Xian, is immense but not yet excavated, appearing in the landscape as a small mountain. Nearby, Qin's famous pottery soldiers have been unearthed. Chou and his development team based the game on what is known about Qin's elaborate ideas and plans for the tomb.

Screen shots of Luyen Chou's edutainment adventure *Qin: Tomb of the Middle Kingdom.*

For instance, the emperor had his architects build elaborate booby traps such as armed cross bows, fire pits, and a pool of mercury (all hazards in the CD-ROM) and then killed the builders and designers by having them sealed into the tomb.

The tomb was also to contain elaborate palaces, gardens, and treasure rooms, all of which appear in the CD. Players navigate around inside the vast tomb, making astonishing discoveries while trying to avoid the booby traps. As you play, you learn more than a little about Chinese history, culture, and language, quickly accessible in the game's encyclopedias. The more you learn, the more fluent you become at solving conundrums.

Chou describes his educational philosophy this way:

> Can you make learning something so compelling it becomes something *you* (the student) really want to do? It's beyond entertainment. It's not a matter of glitzy graphics, sound, or animation. It's more a matter of what I call "the compelling quest-ion."
>
> I use quest-ion in hyphenated form because it's the notion of the "quest" that's at the heart of any good game. It's a quest for some kind of holy grail. True entertainment like true education is question-based.

I try to embed that into every product that we have. And you try to give [players/students] the tools to answer their questions.

Although the *Qin* project and LTI's other products are all designed for CD-ROM, Chou is keeping a sharp eye on developments on the Internet and Web. He is particularly interested in the communications capabilities of the Internet, although he feels the Web still has a way to go before it is a commercial venue for multimedia. He's particularly excited about hybrid products that synergize the strengths of CD-ROM and the Web:

> In something like networked Doom [a multiplayer action game] they're using the local CD-ROM disc as the source of the graphics and the bandwidth intensive stuff and using the Net for the actual event communications amongst players. I think that's going to be huge.

In games like networked Doom, all the content, the assets, and the graphics for the other players are present on each local machine, courtesy of the CD-ROM. The network communicates events. So when you move forward, a signal is sent out over the network to all the other players: "Player 1 moved forward." This is happening at lightning speed, and the other players' machines update their graphics appropriately. So the net enables the interactivity between players, and the CD-ROM provides the high-resolution look and feel of the graphics.

Chou feels that the experience of playing with (or against) other humans is far superior to merely playing against your computer, and he already has plans for a networked version of the *Qin* CD-ROM:

> Eventually what will happen is that when you want help, you will go into our Web database. You'll access it as part of the game's encyclopedia. We'll have a discussion group that allows you to obtain clues and contact other explorers. Beyond that you'll actually encounter other explorers in the site who are also doing the same thing you are, but they will really be there via the Internet.

The large team that developed *Qin* is not atypical of what it takes to produce a complex original edutainment title: a project manager, technical producer, lead engineer, interaction designer, interface designer, art director, prototype programmers, content developers, 2-D and 3-D artists. All are watched over by Chou as executive producer. The team is spread across the country, with members in Dallas and San Francisco in addition to the business and editorial offices in New York. There's even a programming group in Bulgaria. And what do they do to stay in touch? They play computer games via the Internet, after hours. And they're not snooty about the types of games they play, which tend toward the "enter" side of tainment. A current favorite is *Marathon*, which involves destroying vile alien monsters before they do unto you. Says Chou, who has been known to join the game sessions: "All day we're cooperating deeply, trying to build the educational games of tomorrow, and after hours we're killing each other on a virtual battlefield. They go together."

When Chou speaks about the World Wide Web, a slight wistful tone comes into his voice. A couple of years ago CD-ROM development was the height of "cool" on the content side of the bit business. Even though the Web can't begin to match the quality and speed of CD-ROM in delivering complex multimedia, it's clear that excitement surrounding the rise of the Web has stolen some of the glory from other forms of new media.

All over America digital entrepreneurs, from Luyen Chou to Bill Gates, are shaking their heads in astonishment at the same time as their palms move reflexively together in anticipation of the phenomenal World Wide Web.

The Wild World Web

> *The World Wide Web is like a library card catalog scattered across the floor.*
> Attributed to Mitch Kapor, founder of Lotus Development Corporation and the Electronic Frontier Foundation

Is the World Wide Web a new medium? Or is it a flashy, super user friendly incarnation of the older, stodgier, and more arcane Internet. Whatever. It has quickly captured the imaginations of millions and grows thousands of new sites every week. Let's call it a new medium.

The mundane to the incredible and back again are only a click away when you are cruising the Web. Cat lovers have a site where owners can post photos of their pets. One entrepreneur has created an on-line auction house. A manufacturer of earth-moving equipment has a site containing his catalog, complete with downloadable video clips of his machines in action. And each day hundreds of companies are putting their catalogs, product databases, marketing materials, and contact information onto the Web. You can play *Stellar Crisis* and pit your wits and materiel against the intergalactic forces of dozens of on-line opponents for control of the universe. There's even a "virtual cemetery" where family and friends of the deceased can memorialize their loved ones with text, photos, voice, music, and video.

True to the egalitarian, decentralized, nonhierarchical (some use the term "anarchic") nature of the Internet, anyone or any entity can gain a presence on the Web—

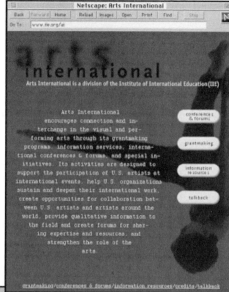

Revolutionary new publishing paradigm or "card catalog spilled on the floor"? Two among the millions of sites that exist "side by side" on the World Wide Web.

spending little or nothing to quite a chunk of change, depending on your budget and ambitions. The resumé or family photo gallery of John Smith of Smalltown, Iowa, exists side by side with the corporate message and expensive screen graphics of a Fortune 100 company or a successful Webzine, which receives thousands of dollars of corporate ad revenues. Depending on your mood, John Smith's page, with its wacky family album, and perhaps his treatise on how to improve the world, may be the more interesting visit. If you're so moved, you probably can click on his e-mail address and dash off a quick note to him.

But the same thing that motivates a John Smith also motivates the Fortune 100 company: the fact that millions of people and other businesses are on-line and only a click away from a visit to your page.

Right now the Web is a good conduit for text and low-resolution photos and graphics. Bandwidth and slow modem speeds act as a bottleneck on sound, music, and video files. Although many sites offer downloadable media, waiting 10 minutes for a short QuickTime video clip may be more of a nuisance than edification. CD-ROM is clearly still the medium of choice for complex interactive applications containing sound and video. Some companies have developed clever ways to combine the relative bandwidth strengths of CD-ROM with the communications strengths of the Internet and Web. And thousands of new and not-so-new companies are tumbling over each other to improve the bandwidth, functionality, and content of the Web. Many envision a more fully evolved version of the Web, eventually offering all the interactive capabilities of CD-ROM, and more.

The Hyperlord of Xanadu

One of the spiritual and technical parents of the Web is Ted Nelson, who, inspired by the work of Doug Engelbart, developed the concept of hypertext and hypermedia more than 25 years ago. Hypertext is, of course, the prime navigational metaphor for the Web. You click on a highlighted word or object, and you quickly (depending on your modem speed) "go" (or are sent) to a new page or site. Anything and everything on the Web can be hyperlinked. If you want to refer readers of your Web page to text or other media on the Web, you don't need to copy it and reproduce it at your site. You simply make your mention of it "hot"—highlighted and programmed with the Web address (called a Universal Resource Locator, or URL) of the material that interests you. Likewise, other sites can include hot links to your pages. This virtually infinite capacity for crosslinking is a key reason the World Wide Web is such a powerful, innovative publishing medium.

Xanadu

Well before the advent of the Web, Nelson and his colleagues began to nurture an even more advanced idea baby—named Xanadu. One of the thorniest and ubiquitous issues in cyberspace is the problem of copyright—achieving proper and appropriate recognition and remuneration to creators or owners of original writing, photography, artwork, music, video, software programs, and any other form of media. Without copyright protections, what gets published on the Web will always be limited because few writers or publishers can work for free forever. And yet, if the cost of copyrighted material is prohibitively high, either the copyright or the material will be ignored. My use of a particular photograph in a document I'm producing may have a value of only 10 cents (or no value, until someone reads my work and pays me), but because of the costs of processing my request and the overhead attached, the image bank charges a minimum of $75, with no exceptions. So, I either pirate the image and hope I don't get caught, or I don't use it even though it was perfect for my project.

Xanadu is a vast and grand scheme to end the copyright bottleneck, revolutionize publishing, and even greatly encourage the legal, paid-for, use of copyrighted materials by anyone for any purpose under the sun. Under the Nelson system, all works would have attached an invisible digital code. Xanadu would automatically meter and keep track of all coded works and handle all copyright transactions.

Here's an example of how it might work:

- I'm making an original multimedia artwork, but I use several images from a favorite photographer, quote a poet, and use a short video clip.
- I don't have to ask permission, nor does another human have to ponder my request and figure out what to charge and communicate all of this to me with a bill. Xanadu is watching and takes care of all these matters.
- I make my work available via the Web. So people can decide if they want it, I publish a portion or selections with free access to all. When users want the whole work, they register their code, perhaps by simply clicking a button on their browser.

- People begin to read and use my work for their own entertainment. They also "quote" from it in projects of their own.
- With each use, Xanadu debits a small sum (perhaps even fractions of a cent—Nelson's "nanobucks") from the user's account and credits mine with some of that transaction—as well as the publishers or artists I used in my work.
- The code (identifying the author and his or her account) for each discrete piece of original work stays attached to the work "forever." Even something as small as a five-word maxim could theoretically be tracked.
- Because the "paperwork" is done automatically and electronically for all transactions, there is virtually no overhead, and even fractions of a cent can be processed practically.
- In a market like the World Wide Web, where there are millions of potential readers and users, an individual author or publisher might do very well even though the royalty from any single use is minuscule.

Some critics of such a global publishing system don't like the process because they believe with long-time info guru Stewart Brand (founder of *The Whole Earth Catalog* and author of *The Media Lab*) that "information wants to be free." And not only wants to but *should* be. Others dislike global publishing because it removes all control over how and by whom copyrighted material is used. Mainstream publishers may be uneasy with the scheme because it would seem to decentralize the whole publishing process. It's clear, though, that for the information superhighway to support commerce and prosper, something like Xanadu will be crucial.

The Web's the Thing

Some observers see the Web as a temporary phenomenon that will somehow be replaced eventually by a commercial network. But certain futurists like George Gilder see it evolving into an extremely formidable entity that will restructure the whole business of PC's and software, perhaps even challenging the hegemony of the megacorporations like Microsoft. As people become comfortable doing business over the Net

Cyberspace and Publishing: Who Owns What?

Esther Dyson has thought and written a lot about copyright, intellectual property, and implications of the digital revolution in publishing. Dyson is the publisher of *Release 1.0*, a $595 per annum newsletter that makes its way to some 1,600 computer industry titans, many of whom participate in her yearly gathering *PC Forum*. Dyson also serves as chair of the Electronic Frontier Foundation, dedicated to protecting civil liberties in cyberspace, and heads a foundation that supports hi-tech ventures in Eastern Europe and Russia.

In a recent interview in the *New York Times Magazine* and in an article in *WIRED*, Dyson explains some of her thinking about the effect of cyberspace on publishing: "The problem for providers of intellectual property in the future is this. . . . They will operate in an increasingly competitive marketplace where much of the intellectual property is distributed free and suppliers explode in number.

"Economics would say that since the supply of content is increasing, the costs of duplication and distribution are diminishing and people have the same amount of time or less, we are all going to pay less."

In her vision this means that creators and publishers of intellectual property from fiction to software will have to get clever about collateral forms of income—like personalized support for complex software and public appearances by authors.

Creators and publishers of unique or highly valued work like best-selling novels or popular electronic games may find protection through some kind of metering scheme like Xanadu. But the more

using encrypted credit card numbers, many interesting capabilities arise. People may demand and buy only certain features of a software program (like Windows) as they need them. And they may simply rent the use, on a metered basis, of programs they seldom use. After all, if and

Esther Dyson has been a long-time commentator and major influence in the world of computers and communications.

generic kinds of content, like news and financial data, she feels, will be part of a scramble to find ways to make it pay. This doesn't mean anarchy. Dyson writes: "This new world will distribute its benefits differently than how they are distributed today, but as long as the rules are the same for everyone—and predictable—the game is fair. The big issue is the transition.

"The most promising businesses in the Net world will be services and processes. They will include selecting, classifying, rating, interpreting, and customizing content for specific customer needs. Other services will include access to various sorts of performing, interacting with people, and all kinds of other activities that require the time of a live, talented person."

Even though Dyson spends most of her professional life thinking, writing, and talking about the details and the big picture of the digital revolution, she's not fanatical about it. Says Dyson:

"People should not be made to feel socially inadequate if they are not wired. The important thing to remember is that this is not a new form of life. It is just a new activity."

when software, information, and media are all instantly accessible online, there's no need to have them sitting idly on your own hard drive. Gilder forecasts this phenomenon as the "hollowing out" of the PC. If this happens, the insatiable "Blob" will have its comeuppance.

Digital Misgivings

Gilder is definitely not alone in his technological optimism. As thousands of young companies have started their wagons for the new information frontier, they are accompanied by an onslaught of articles, books, and TV specials full of up-beat heraldry for the digital revolution and new media. Over the past year or two, though, another phenomenon has appeared more and more frequently in the press. This is the worry piece that ranges from stern criticism to near panic.

And what are people worried about?

Equity

A number of critics and even some new media enthusiasts worry about haves and have nots in the new digital society. After all, the equipment is expensive, and the software tends to be arcane and hard to use. Are people with no money and little education to be left completely behind? And if they do gain access to equipment and the ability to use it, what is really there for the underclass on the Internet and in digital new media?

In a world in which nearly every transaction is mediated by computers and networks, what happens to those who don't have access?

One answer may be that it is just a matter of time. In a society in which computer users are still a minority (albeit a growing one), it comes as no surprise that the poor are in the nonuser group. The telephone began as a curiosity and plaything for hobbyists, rich men, and professionals, like doctors. Over the course of a century, and with the help of AT&T's goal of universal telephone service, and with the increasing perception of the phone as a necessity, telephones spread to all economic groups. Offering reduced rates to people on welfare or disability has also aided diffusion.

Access to networks may follow a similar pattern if, as appears likely, being "wired" comes to be seen as a necessity. The advent of the so-called Web PC's (also called Internet appliances, browser boxes, or nettops)—inexpensive workstations that function only as connections to the Net and the Web—could also aid the process of diffusion.

Not content with a laissez-faire approach, activists have leaped in to provide computer and network access and training to citizens of the inner city and other disadvantaged communities. Web resource pages for every ethnicity and social interest group in America have already been constructed, and new resources appear daily. The fact that the Telecommunications Act of 1996 requires that telecom companies provide services to schools and other public institutions at discounted rates should also facilitate equity in access.

Crime, Ethics, and Privacy

Another high-profile worry is the issue of crime on the Internet and threats to privacy. Real-world crimes, like breaking and entering, confidence games, terrorism, larceny, vandalism, and peeping tomism, all have analogs in cyberspace. The word "hacker" was formerly a term of affection and even respect for the devoted computer geek who regularly stayed up all night getting computers to do things no one thought they could. But the persistent efforts of a number of clever teenagers and even a few real-world international spies have made "hacker" a synonym for thievery, mischief, and mayhem.

Clifford Stoll, in his 1989 book *The Cuckoo's Egg*, documents in great detail his frustrating but ultimately successful sleuthing job that ultimately netted a mysterious hacker who was breaking into the nation's military and research computers. The culprits turned out to be two West German hackers in the pay of a KGB operative in East Berlin.

Perhaps the most famous hacker on our own shores is Kevin Mitnick, who obsessively roamed the Net for 15 years, breaking into FBI and DMV computers and private e-mail accounts, poking around and occasionally pirating files and software. At one point he even obtained thousands of credit card numbers, though he apparently never used any. Even more effective than his hacking skills were his confidence ruses, and he successfully pried all manner of information from phone company employees by impersonating superiors and even sifting through trash to obtain useful intelligence.

When the *New York Times* reporter John Markoff chronicled his activities in 1991 in the book *Cyberpunks*, Mitnick retaliated by hacking into the reporter's e-mail account to eavesdrop and then boasting about it. He finally met his match in 1994 when he broke into the computer of

Tsutomu Shimomura, a security expert with even greater hacking skills than Mitnick, to obtain some special cellular phone software. Mitnick was envisioning performing his hacks via stolen cellular accounts as he moved around the country—making himself virtually uncatchable. Shimomura did catch him, though, and published a serial account of it in *WIRED* magazine, later published as the dramatically titled book *Takedown: The Pursuit and Capture of America's Most Wanted Computer Outlaw—By the Man Who Did It*.

Another high-profile issue is the ethics of Net use and the problem of incivility—the gratuitous, extreme rudeness and boorishness known as "flaming." Youthful computer nerds and others, perhaps reveling in the relative anonymity of on-line and the fact that they can "communicate" with people they are unlikely to ever meet, cut loose of the moorings of good manners to spew invective and threats to anyone who annoys them.

People who have been flamed or harassed on-line report it a very unsettling experience. James Gleick, science writer and columnist for the *New Yorker* and the *New York Times Magazine,* has had the "flaming" experience:

> We have reached a time when the on-line world is more than just a mirror or an extension of the real world. Cyberspace is as much a frontier as the New World was to seventeenth-century Europe. That's nothing to romanticize. Frontier worlds are known to be rude ugly, and lawless. Too often, there is a degenerative "Lord of the Flies" mood. People lie shamelessly, and other people believe them. Angry teenagers screech at one another behind pseudonymous masks. I've had some shockingly unpleasant electronic encounters.... Just the other day, a disgruntled young man whom I've never met posted a public message expressing the desire to see my hands blown off in an explosion.

The on-line services, such as Compuserve and America Online, have rules about civility, obscenity, and hateful language that they can enforce by banning an out-of-control, abusive member. On the wild, wild Internet, enforcement of civility is a more difficult proposition. Software, sometimes called a "bozo filter," exists to automatically delete messages from a particular source. This is some help, although the "bozo" can also get around it by obtaining a new address and on-line identity.

John Kaufman, a writer based in San Francisco, had an unsettling variant on the bozo experience when he attracted the attention of a cyberstalker. Kaufman was a regular contributor of "posts" to a variety of the some 15,000 Usenet groups on the Internet. At first he thought he'd gained a fan when a woman e-mailed him to tell him she admired his work. He grew apprehensive when she began sending him daily e-mail commenting on a range of his postings from Latin American politics to remarks about the weather in the Shetland Islands. When she sent him a three-page letter that seemed to contain a complete dossier on his whole life, from his mother's occupation as a concert pianist to the name of his cat, he was astonished. It turned out that the woman had used some of the powerful new search programs, including one called *Deja News*, that can quickly sift through hundreds of megabytes of Internet postings to supply an "author profile" on a given person. *Deja News* supplies the requester with a list of every posting or "article" that person has made. The stalker had used the list to compile her dossier on Kaufman.

The Telecom Act of 1996 contains a provision that makes it a crime to "repeatedly initiate communication . . . solely to harass any person," which may help curtail some flaming and stalking.

Pornography

Pornography on the Internet is probably the biggest issue in the news and, by no coincidence, also in Congress. There are various sources of porn from BBS's to Usenet groups. Pornography on-line generally involves the same issues that it does off line: What is it? (How is it defined?) Should it be controlled? If so, how and by whom? Do laws against it abridge the free speech guarantees of the First Amendment?

Free speech advocates argue against regulation, making the point that porn is no more a problem on the Internet than it is in any other publications medium. Parents and teachers, they argue, can take measures to prevent access.

Proponents of regulation argue that the Internet is different. Any one of any age with the savvy, prurient interest and a modem can access pornographic materials, which would be much more difficult for them to obtain in the world of book and video stores.

On-line entrepreneurs have responded to the problem with a variety of options for parents and teachers. Software is available that allows a parent to block access to any site or address on the Internet or Web, such as anything with the suffix .sex. The software can also prevent children from giving out their home phone numbers, addresses, and the family's credit card numbers and can even limit the amount of time kids can spend on-line. It can also keep a record of all programs used and sites visited by the child.

Federal authorities have also become active on-line, conducting a variety of surveillance and sting operations, including one in September 1995 that resulted in arrests of more than a dozen subscribers to America Online on charges of purveying child pornography. The Communications Decency Act passed with the Telecommunications Bill of 1996 attempted to ban all pornography on the Internet, but was considered overly sweeping by many free speech organizations. The act was overturned in an appellate court decision, which strongly upheld full First Amendment protections for the Internet.

The facts of life are such that the Internet increasingly reflects every aspect of modern society, from the creation of hopeful new communities to the full range of depravities, criminal activities, and the law-enforcement techniques needed to combat them.

Living Digitally

The Internet has become the most highly perfected means yet for the scattering of the self beyond recall. Unless we can recollect ourselves in the presence of our intelligent artifacts, we have no future.

> Stephen Talbott, *The Future Does Not Compute: Transcending the Machines in Our Midst*

Electronic man wears his brain outside his skull and his nervous system on top of his skin. Such a creature is ill-tempered, eschewing overt violence. He is like an exposed spider squatting in a thrumming web, resonating with all other webs. But he is not flesh and blood: he is an item in a data bank, ephemeral, easily forgotten, and resentful of that fact. . . . When we lose nature as a direct experience, we lose a balance wheel, the touchstone of natural law.

> Marshall McLuhan and Bruce Powers, *The Global Village*

There has been much written both celebrating and denouncing cyberspace, but to me this seems a development of such magnitude that trying to characterize it as a good thing or a bad thing trivializes it considerably. I also don't think it's a matter about which we have much choice. It is coming, whether we like it or not.

John Perry Barlow, 1995

Perhaps the most interesting and profound of digital misgivings has to do with issues of mental health, social health, and our near- and long-term social destiny. What are the personal affects of a large diet of computer-mediated communication? Beyond concerns about electromagnetic radiation, eyestrain, and sedentariness, what does it do to the mind and world view of the user? What are the affects on society of large numbers of people interacting via new media and networked computers?

Are we losing our selves and our souls in a vast electronic fiction and dream world to the detriment of "real" life and the exigencies of our biology, our history, and our destiny? Is the World Wide Web a more apt name than was intended? Is it an extension of the web of life, or perhaps an evolutionary dead end and death trap?

John Perry Barlow.

Harper's Magazine, in a roundtable debate entitled "What Are We Doing On-line?," pitted two of the most prominent and articulate of the info gurus against two equally articulate digital curmudgeons to discuss some of the above concerns. The cyberspace proponents were Kevin Kelly, author of the book *Out of Control* and editor of *WIRED* magazine, and John Perry Barlow, one of the most well-known netizens and co-founder of Electronic Frontier Foundation, a group formed to protect civil liberties on the Internet. The digital doubters were Sven Birkerts, author of *The Gutenberg Ele-*

Kevin Kelly, Executive Editor of *Wired* magazine.

gies: *The Fate of Reading in an Electronic Age*, and Mark Slouka, author of *War of the Worlds: Cyberspace and the Hi Tech Assault on Reality*.

Although the two sides end up agreeing to disagree, they each present persuasive arguments. Birkerts expresses resentment over the idea that, as Barlow puts it, the digital revolution "is coming whether we like it or not." Birkerts disagrees: "We are being forced to adapt by a pressing social consensus that seems to say that if you don't have 'x' you're out of the loop.... We're looking to technology to solve what it has wrought.... The last two words in my book are 'Refuse it.'"

The "mediation, mediation, mediation" that Birkerts decries (in the quote at the beginning of this chapter) is in opposition to what he values: "What seems most important to me is focus, a lack of distraction—an environment that engenders a sustained and growing awareness of place and face-to-face interaction with other people."

While supporting his values, Barlow and Kelly question his "conservative" nostalgia for the good old days and what they suggest is an overly sentimental regard for the Book. Says Kelly: "Where am I when I am involved in a book? Well here's the real answer: you're in cyberspace.... You think that somehow a book is the height of human achievement. It is not."

Mark Slouka views the furious interest in cyberspace as a retreat from an increasingly toxic reality:

> The wired world is a response to certain cultural changes over the last two or three generations—the breakup of the family, the breakdown of the community, the degradation of the physical environment.... Every place I've loved

in this world has been paved over, malled over, disappeared. As we observe this assault on the physical world, we feel ourselves losing control. I think alternative worlds become more appealing to us.

John Perry Barlow responds to both Slouka's and Birkerts' concerns with a "no pain, no gain" credo:

> I spent seventeen years driving a four-horse team around, living in direct contact with the phenomenal world and my neighbors. And what I finally concluded was that there were so many forces afoot that were in opposition to that way of life that the only way around technology was through it. I took faith in the idea that, on the other side of this info-desert we all seemed to be crossing, technology might restore what it was destroying. There's a big difference between information and experience. . . . But if we're going to get back into an experiential world that has substance and form and meaning, we're going to have to go through information to get there.

These sorts of arguments will continue. There is no question that we are treading in a strange new environment that can be alienating and dislocating. The world of print and the book seem normal and comfortable because they are familiar. Looked at from an outsider's point of view, though, the act of reading is strange indeed. Who could have imagined that by scanning lines of black marks on a page we could have opened to us whole systems of knowledge and wisdom? That fictional characters sprung from the imaginations of great writers would become through the act of reading nearly as real to us as flesh and blood humans?

There will probably rise up communities of people who decide, in Sven Birkerts words, to "refuse it" and adopt an Amish or halcyon way of life, eschewing all modern telecommunications. Those of us who decide to "go through it" will not be able to abandon the power of the textual word. But we will increasingly learn, educate ourselves, and communicate via machines that we have empowered to simulate features of human intellect.

CHAPTER SIX

The AI Wildcard

Words also and thought as rapid as air
He fashions to his good use . . .
                Sophocles, *Antigone*

The giant centralized computer with its whirring tapes and complex cooling systems will be supplemented by myriad chips of intelligence, embedded in one form or another in every home, hospital, hotel, every vehicle and appliance, virtually every building brick. The electronic environment will literally converse with us.
                Alvin Toffler and Heidi Toffler, *The Third Wave*, 1980

I believe that current programming languages are the larval form of something far more interesting that will mature in the next 10 years—a new form of communication on the same level as speaking and writing.
                Jaron Lanier, 1990

One of the most ambitious science efforts since the end of World War II has been the quest for artificial intelligence (AI), sometimes called machine intelligence. The goal of AI differs according to who is defining it and how confident he or she happens to be feeling. At its strongest, though, the goal is nothing less than to finesse millions of years of evolution and create an autonomous intellect in a computer equivalent or superior in every way to human intellect—including problem solving, common sense, intuition, creativity, and not last or least communication and language abilities.

John McCarthy, director of AI work at Stanford University, offered a subtly toned down version of the mission, defining AI as "the science of making computers do things which if done by men would require intelligence."

Spurred by promising early research, two seminal AI thinkers, Alan Newell and Herbert Simon of the Rand Corporation and Carnegie Mellon University, predicted in 1957 that in ten years a computer would be world chess champion and create musical compositions of "genius." In 1967 Marvin Minsky of MIT, the dean of classical AI theory and research and perhaps its most unabashed proponent, proclaimed: "Within a generation . . . few compartments of intellect will remain outside the machine's realm—the problem of creating artificial intelligence will be substantially solved."

If AI is successful in bringing the "compartment of intellect" that mediates language, meaning, and communication into the realm of the machine, it would have profound implications for every aspect of our society, beginning with how we "communicate" with our computers (and ultimately all other machines in our lives) to how we create and utilize all new media (see *Future Bright* for two scenarios). As the Tofflers wrote in their best seller *The Third Wave*, the environment would "literally converse with us." Reaching such a goal, as we shall see, or even bringing it within sight has been anything but easy and straightforward. But the process of trying reveals more and more about the complex nature of language and how we use it to communicate. Over the years the AI effort has spun off a number of important disciplines, including the burgeoning field of computer-mediated communication (CMC), which has received a real boost from the rise of the Net and the Web.

Believers and Skeptics

Early statements and claims by the AI gurus raised the eyebrows, incredulity, and ire of a number of other thinkers. To them, these boasts looked like the hubris and arrogance of a sort of twentieth-century alchemy. With millions of dollars of research patronage at stake, they were promising to take the lead of 1's and 0's in integrated circuits and transmute it into the gold of intellect, the crown of creation. An acrimonious debate sprang up between the strongest promoters and believers and these very articulate doubters and critics of the whole premise of the AI project. The debate reached its highest pitch in the mid-1980s.

In a number of monographs and books, including *Alchemy and Artificial Intelligence*, *What Computers Can't Do*, and *Mind Over Machine*, Hubert Dreyfus, a philosopher at the University of California Berkeley, questioned nearly every aspect of the AI endeavor—from its philosophical premises to its claims of results and imminent success: "Despite what you may have read in magazines and newspapers . . . twenty-five years of artificial intelligence research has lived up to very few of its promises and has failed to yield any evidence that it ever will." Dreyfus and others argued that without a body and all that comes with it—the universal experiences like pleasure and pain, feeding, growing up, love, loss, anger, jealousy, and all the vast realm of being alive in the world—a digital computer hadn't a prayer of developing intelligence and becoming a great communicator. He also argued that what the mind does is not reducible to the sets of rules and heuristics and algorithms used in software programming. He predicted that, like a man climbing a tree to reach the moon, AI would experience encouraging results at the very beginning that would soon peter out in the uppermost branches. If a computer performs intelligent functions, it is a reflection of the many human intelligences that engineered the hardware components and the layers of software languages—not a sign of some independent "intellect" in the computer.

In fact, forty years after the Newell-Simon prediction there is still no computer chess champion, although IBM's Deep Blue is certainly knocking on the door. But Deep Blue doesn't play chess the way humans do. It uses what's called the "brute force" approach, utilizing its powerful computing engine to analyze up to 50 billion moves in three minutes of "thinking." The "musical compositions of genius" from a computer are nowhere in sight, nor have the problems of creating AI been solved. The goal is as elusive now as it was 30 years ago, but there is something deeply engaging in the ambition. Humans have long been fascinated by automata like robots and simulacra that model intelligent human behavior. (A few decades after Mary Shelley wrote *Frankenstein*, Alexander Graham Bell and his brother were using cadavers to investigate the nature of speech and hearing.)

The effort to make computers do intelligent things has gone hand-in-hand with making them easier to use and more useful as communications devices. And the AI project has attracted brilliant computer scientists who have developed some of the most sophisticated software programs ever attempted, with a variety of spin-off applications that have profoundly affected the nature of new media and communications.

Classical AI research of the past several decades has concerned the following fields:

- Robotics—endowing mobile machines with abilities like finding their way through a maze, and perhaps even finding and plugging themselves into a wall socket for a recharge when their batteries run low. Uncannily sage to hopelessly stupid, all in the same machine. The mechanical servant that follows you around and does your bidding, like C3PO or R2D2 of *Star Wars* fame, is nowhere in sight.
- Machine vision—endowing a computer with the ability to "see" objects and identify them via video cameras, pattern recognition, and analysis programs. The ability to separate an object from its background and identify the object from any perspective are hallmarks of human vision acquired at an early age. But they are immensely difficult tasks for a computer to do at all, much less consistently. Being able to recognize unspoken components of communication, like gesture and facial expression, could, of course, help computers make sense of speech.
- Expert systems—these programs attempt to model areas of human expertise—particularly in complex information-heavy fields like medical diagnosis. For instance, a doctor can enter a description of a patient's symptoms, and the program uses its database and a wide range of rules of thumb (called heuristics) to present the doctor with a list of possible diseases, a probability estimate for each, and the rationale for its diagnosis. Researchers employ some of the techniques used to create expert systems in trying to develop intelligent agents.
- Speech synthesis, speech recognition, and natural language understanding: Because of their potential impact on communications, the rest of the chapter focuses on these technologies.

The Multibillion Dollar Holy Grail

A major goal of AI is to make our interactions with computers more natural and intuitive. Being able to converse with them by talking and listening is a vital part of that process.

Raymond Kurzweil, *The Age of Intelligent Machines*

We're all thoroughly inured to the string of ads, future scenarios, and science fiction series that depict humans pattering away to their computers and

Future Bright

In 1988, Apple Computer produced an attractive little filmed vignette called *Knowledge Navigator*, depicting the company's vision for future use of networked computers. In the film, a professor steps into his well-appointed home office and begins a dialog with his computer lying on his uncluttered desk. He conducts most of his business through his "intelligent agent," an AI program that appears as a talking head on screen, a very polite and jaunty youth in a bow tie. The professor is preparing materials for his lecture later that day, and in the course of a few untroubled minutes makes an important discovery about the rain forests of Brazil by touching the screen and asking his agent to call up various charts and diagrams. He also has the friendly IA get a colleague on the videophone and persuades her to make a "guest appearance" at his lecture.

The whole vignette leaves the viewer coveting this "around the corner" technology and fully persuaded that Apple's computers and communications of tomorrow will be charmingly integrated into our lives, having completely removed the bafflement and stress from the information explosion.

Connections: AT&T's Vision of the Future goes Apple two or three better. Produced in 1993, the film purports to gives us a slice of life in 2013. Luckily, all of the intractable problems of telecommunications, AI, computer understanding, virtual reality, and speech recognition have been fully solved. Courtesy of AT&T, there is bandwidth to burn. It's a fully "mediated" world with high-definition, networked screens everywhere—whether as hand-held PC's or on the back of every seat of an airplane or as wall-sized flat screens in every room and office.

In addition to their ubiquitous and spectacular communications systems, the people are successful, wealthy, beautiful, well-dressed, and very nice.

- The husband, an architect, has a "dynabook" that offers full

3-D building renderings in the field and features a cellular videophone.
- Desktop systems feature screens that are thin, transparent slabs of lucite with holographic capabilities. "Sidney," the doctor/wife's Intelligent Agent, is visible in the screen from any angle and—always ready to be of assistance—turns to follow the motions of his mistress (or whatever you call the owner of an IA) as she moves about the room.
- At one point when his wife is out of the room, the husband goes over to Sidney and asks him why his wife designed such a young, handsome IA for herself. And Sidney replies, with perhaps just a slight smirk, "I am not authorized to tell you that. . . . But thanks for the compliment."
- Techno wonders multiply: There's a scene in which daughter and mother go shopping for a wedding gown by sitting in Mom's office and calling up their boutique on the immense wall screen. Daughter's registered "mannequin," a perfect, fully mobile 3-D likeness of herself, models various gowns, which can be instantly altered on screen at the spoken request of mother or daughter.
- In another scene, Mom—always with Sidney's well-mannered help—sets up and performs an on-screen consultation with a young woman patient who plays World Cup soccer with a prosthetic leg. Images of the prosthesis in 3-D come and go on screen. Mom's colleague, who's on a fishing trip, sends his IA to participate in the consultation as a proxy.
- Flawless automatic instant machine translation is available for any international call. When the father can't understand his daughter's Belgian fiancé during a video call, dad presses the handy "Translation" button on the pay phone screen and *Voila!* it transforms into idiomatic English somehow delivered in the fiancé's voice.

It's a world that's sleek and smooth, and technology finally really does "make life easier and more rewarding for people" (from the package blurb)—and not just a little spooky. (Do intelli-

> gent agents gossip about their owners when they're off duty? Can you tell them to stop looking at you all the time without hurting their feelings? What if bad guys get a hold of your "mannequin" program and use it to scam your friends and family?)
> Connections is the Wonderful World of Tomorrow served up the way we like it, a kind of optimistic American theology of the future that has been a part of our scenarios throughout the twentieth century. Whether these types of scenarios benefit us by giving us something bright and hopeful to look forward to or whether they contribute to a perpetual disappointment and dashed expectations about what technology can do for us—or both—they are a fixture of our imagined futures.

receiving appropriate, intelligent, helpful replies. The scenario is so common that it's no surprise that many people assume the technology already exists or is "just around the corner"—to echo what certain AI researchers and telecommunications executives have been claiming for years.

In fact, though, even the advent of the word processor that takes dictation (which does not require understanding on the part of the computer) has proved to be a profoundly difficult problem. IBM and other companies have recently released dictation software products, but they come with a variety of limitations, not the least of which is that they must be trained to the individual user. And the much bigger goal of creating a computer with full, intelligent "understanding" of human speech remains an elusive multibillion dollar holy grail that would, of course, revolutionize the way we operate and communicate with every machine on the planet.

Here, though, are some of the reasons that, in spite of what you may read by way of prognostication, you may not want to hold your breath while waiting.

Speech Synthesis

Let's start with a technology that has been largely "achieved": speech synthesis, which enables your computer to read aloud to you. Linguists have long had programs that reproduce the sounds of specific phonemes—the units of sound that make up human speech—and put them together to make understandable words. The computer can "read" any ASCII text to form the consonants, vowels, syllables, and words. It can likewise read punctuation to create appropriate pauses and stops. Your computer, of course, does this with no understanding of what's being spoken, and although you can choose between a number of voices (male, female, etc.), the speech is flat and expressionless—very "machinelike." There's a project under way at MIT's Media Lab to add more "expression" to voice synthesis, but just how lifelike the speech will be without the computer having any sense of the meaning will be interesting to hear.

For people with normal sight, speech synthesis is an amusing novelty and not of much practical use. There *are* applications that enable you to call into your computer and have text, like e-mail, read to you over the phone—if, say, you were delayed at the airport and couldn't access it via your laptop. But speech synthesis becomes a vital tool for blind users or those who have lost the ability to speak. For them, synthesis can become the prime interface with their computer (and a crucial link to other people), with the program automatically reading text aloud in the currently active window. The computer can also become a "voice" for people who have lost their own. The British astrophysicist Stephen Hawking, severely disabled by Lou Gehrig's disease, communicates with a speech synthesizer and even uses it to deliver talks. He jokes that its only drawback is that it gives him "an American accent."

Automatic Speech Recognition (ASR)

Because of its many current and potential applications, a computer that can decode human speech (called automatic speech recognition, or ASR) has been a much sought-after prize.

Phonemes are created by the vocal chords, lips, tongue, teeth, mouth, and nasal cavity. We use about 16 different vowel sounds and 24 consonant sounds to create some 10,000 different syllables in English. Humans typically speak about three words per second with an average of six phonemes per word for a total of 18 phonemes per second. Add to this the fact that we seldom pronounce a word precisely the same way twice (adding dynamics and emphasis according to the context) and that we speak at differing speeds, depending on the situation, our moods, and the time of day—and the fact that speech varies widely according to gender, age, and region—and you can begin to see the scope of the problem.

An even more difficult problem for researchers is the fact that when we speak naturally *weslurourwordstogether* into one long stew of sounds. In part, because humans bring a deep, flexible understanding of meaning and context to conversation, we usually can digest the stew instantly with little apparent effort. As with sight, we also have an interesting ability to sort foreground from background—for instance, the person we're listening to at a party from the noise and other voices. Machines are forced to take other approaches.

The classical approach to machine speech recognition has been to equip the computer with a large database of spectrographs (sound patterns) of many spoken words—called "templates." The computer rapidly analyzes incoming speech for familiar patterns and attempts to match them with the correct template in its database. But this is only the beginning. In his book *The Age of Intelligent Machines,* AI researcher Raymond Kurzweil writes:

> Only a small portion of the technology in a successful ASR system consists of classical pattern-recognition techniques. The bulk of it consists of extensive knowledge about the nature of human speech and language: the shape of the speech sounds and the phonology, syntax and semantics of spoken language.

Kurzweil's company, Kurzweil Applied Intelligence, and several others, including IBM, have released commercial speech-recognition products. By current standards they still use a lot of computing resources. For example IBM's VoiceType requires at least 8 megabytes of dedicated RAM (above and beyond what's needed for the operating system and other applications) and 33 megabytes of free hard-disc space and an additional 30 megabytes during the "enrollment" or training process.

The programs recognize fairly large vocabularies (30,000 to 60,000 words of so-called discrete or dictated speech—which means the user must leave a brief . . . but . . . distinct . . . pause . . . between . . . each . . . word. Using the systems involves a training period of one to several hours wherein the program gets to "know" your voice—and you get used to it. Factors that can seriously degrade the performance of these systems are background noise and changes in the speaker's voice. One reviewer, who began using an ASR system when he developed carpal tunnel syndrome in both hands, was generally delighted to have the ability to work again in

Woman using IBM's VoiceType dictation program.

spite of the program's numerous drawbacks. When he developed a cold, though, the system developed an allergy to understanding him.

For unimpaired computer users, these systems are probably too expensive and quirky to buy and train. With Moore's law in operation, though, it may not be long before discrete speech understanding becomes a standard feature of personal computers and a useful addition to how we get them to do our bidding.

Continuous Speech Recognition (CSR)

Now a computer that could understand normally flowing speech (continuous speech recognition, or CSR)—that would be one big step to the biggest prize of all: a computer that could understand the meaning of language and respond flexibly and appropriately. Both Apple's *Knowledge Navigator* and AT&T's *Connections* scenarios depict a future in which the problems have been fully solved. The million dollar question is, To what extent does CSR depend on natural language understanding in human speech recognition? In other words, does our ability to decode the slur of words in normal speech depend to an important degree on our ability to understand meaning and context through the commonality of human experience? If it does, then are there ways for computers to be programmed to finesse the problem with brute force and clever algorithms as Deep Blue has done in chess? Although there are programs that can recognize and decode whole phrases, the problem is a difficult one. CSR is the big prize because until a computer can decode and recognize normally flowing, nongrammatical language with high fluency and very low error rates, people will choose other ways to interface with computers—unless they are highly motivated because of disabilities, for example.

Natural Language Understanding (NLU)

While it may still be unclear whether or not true CSR requires some degree of human understanding, there is no doubt that getting a computer to understand *meaning* in spoken or typed natural language, which has been a key realm of the classical AI project, goes right to the heart of the debate about what computers can and cannot be made to do. Even when we set aside the project of making sense of speech and restrict the computer to typed-in language, there remain profound problems for machine understanding. Although computers can recognize and offer definitions of any word in the language and parse a sentence faster than your high school English teacher, meaning and understanding remain elusive.

A lot of this has to do with the nature of human languages. Terry Winograd, a linguist and computer scientist at Stanford University, calls language a "blueprint" that humans use to convey and construct meaning. In other words, our use of language to communicate with other humans assumes a broad and deep range of mutual experience as living beings on earth. Although "any child can do it," it has become clear that that's not because it's easy. Human children universally master the complex rules and exceptions of their native language's grammar as well as the meanings of hundreds and soon thousands of words all at the same early age across cultures. This has led some linguists to theorize that language ability is to some important degree genetically determined.

In an interview a decade ago, Winograd confided that the problem of conferring on a computer the kind of deep, flexible understanding of the world that humans have and that is a prerequisite for the ability to create meaning out of language is "at least very very difficult and probably impossible." And in a recent conversation he said that although he no longer works in the field of natural language understanding, he hadn't heard or seen any new developments that had changed his mind.

ETAOIN? SHRDLU!

In the early 1970s, a heady time for classical AI, Winograd became well known in the AI world for creating an especially nifty natural language program called SHRDLU. The program allowed the user to type in normal English requests and directions to the computer regarding a screen full of colored blocks of various shapes and sizes, which the program could move and stack. Winograd and his colleagues had equipped the program with a clever language parser, which analyzed the users' typed sentences, and a certain amount of general knowledge about the BlockWorld. The program could ask for clarification when it encountered an ambiguity or something it didn't understand. This excerpt from a session with SHRDLU gives a taste of what the program could do:

1. pick up a red block
   OK
2. find a block which is taller than the one you are holding and put it into the box
   BY "IT," I ASSUME YOU MEAN THE BLOCK WHICH IS TALLER THAN THE ONE I AM HOLDING
   OK
3. what does the box contain?
   THE BLUE PYRAMID AND THE BLUE BLOCK
4. what is the pyramid supported by?
   THE BOX
5. how many blocks are not in the box?
   FOUR OF THEM
6. is at least one of them narrower than the one which I told you to pick up?
   YES, THE RED CUBE
7. is it supported?
   YES, BY THE TABLE
8. can the table pick up blocks?
   NO

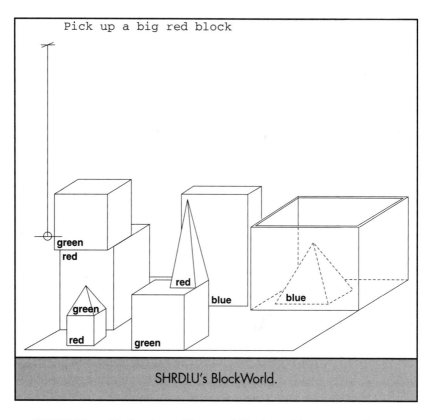

SHRDLU's BlockWorld.

SHRDLU could also learn. If you told it that a blue pyramid on a red cube was called a "steeple," it could from then on build a steeple on command. Because it seemed to intimate that a computer could understand, reason, learn, and exhibit a certain amount of common sense, SHRDLU was a sensation in the AI world, and people briefly thought that true AI had arrived. The program successfully fostered the illusion that you were really conversing with the computer. The problem was that as soon as Winograd and his team tried to expand the program beyond the BlockWorld domain, everything broke down. Winograd later became skeptical of aspects of the AI endeavor and changed his emphasis to more fruitful projects in computer science—what he calls getting computers to "help *us* be more intelligent."

But others kept plugging.

The Psyche of CYC

In what may be either the last hurrah or the Manhattan Project of classical AI's attempt to crack the "understanding" problem, Doug Lenat and his team have been at work for more than a decade on a 25 million dollar project called CYC, as in encyclopedia. CYC is a vast program whose goal is to let computers in on a large chunk of what humans call common knowledge. Lenat, a long-time AI researcher, believes this is a prerequisite or co-requisite for natural language understanding or truly intelligent agents.

He describes AI programs as "idiot savants"—a term confirmed by Terry Winograd's and other researchers' experiences with them. They perform flawlessly in small, specialized areas of knowledge (called "limited domains") like the BlockWorld, but break down the instant they're called on to do anything outside that specialty. As Lenat puts it (in a recent article in *Scientific American*), "Ask a medical program about a rusty old car and it might blithely diagnose measles."

Since the mid-1980s, Lenat's team has been working in Austin, Texas, at the Microelectronics and Computer Technology Group, an R&D effort funded by a consortium of companies like Apple and Xerox. The team has been spoon-feeding CYC—via a computer language called "predicate calculus"—a verbal knowledge of the big, wide world and ways to represent and use that knowledge. Eventually, CYC is to contain some 100 million pieces of knowledge made up of facts and rules for using them.

In the *Scientific American* piece, Lenat describes a few of the problems one little word can pose to CYC:

> An innocuous word such as "in" turns out to have two dozen meanings, each corresponding to a distinct concept. The way in which you, the reader, are in a room is different from the way the air is in the room, the way the carpet is in the room, the way the paint on the walls is in the room, and the way a letter in a desk drawer is in the room. Each way that something can be "in" a place has different implications—the letter can be removed from the room, for instance, whereas the air cannot. Neither the air nor the letter, however, is visible at first glance to someone entering the room.

One can only guess at the problems such usages as *in love, in trouble, in debt, into cross-country skiing,* and *what's in* might cause.

The hope is that CYC will eventually be able to plow through books and magazines, building its own knowledge base, although that ability hasn't materialized yet. And the project has many critics, even in the AI community, where people complain variously that CYC doesn't use enough approaches to modeling knowledge, uses "inadequate logic, and will be buried in static facts in a dynamic world."

It's a heroic effort that may very well be doomed—at least in its original goals. In the article, Lenat opines, "Now that the world has all but given up on the AI dream, I believe that artificial intelligence stands on the brink of success." Whether or not he's right, CYC, like so many assaults on AI before, is likely to have interesting and unexpected spin-offs.

With CYC, like SHRDLU before it, the interface between computer and human is natural language typed in from the keyboard. One of the key projects aimed at making the Internet and Web more navigable and useful is the development of the so-called intelligent agent, a piece of software that knows about your preferences, needs, history, and work habits and can serve as your proxy in finding what you want on the net. The ideal interface with an agent would, of course, be typed-in natural language. Many companies and research groups (see the box "Future Under Construction") are in hot pusuit of some version of this prize. But a truly intelligent agent will need as much sense, knowledge, and understanding of the world as it can muster—or at least simulate—all the kinds of things CYC is trying to accomplish. Most current agents are simply glorified search engines that match the occurrence of keywords and then offer a percentage-based return. If one of CYC's spin-offs helps to make search programs more useful, then CYC may have paid for itself, and AI will once again have spun off something useful to human intelligence even though true machine intelligence is out of sight.

Research goes on. A new generation of scientists has at least temporarily set aside the rigorous rule-based approaches of classical AI in favor of nonrational, "out-of-control," evolutionary models, neural nets, and a burgeoning field called "artificial life," which sets digital automata loose to literally evolve their own directions and solutions.

Perhaps even more interesting than what AI has accomplished is what it wants to accomplish—a companion intelligence in the universe, another communicator. AI ambition and its ambitiousness tell us much about our future because they reveal something important about the directions we dream in—the vectors of our technical desires.

Future Under Construction

Three of the most important centers of computer science and artificial intelligence research over the past 25 years have been the Massachusetts Institute of Technology, Carnegie Mellon University, and Stanford University. Recently, all three programs have sponsored vigorous projects in new media and computer-mediated communications. A quick look at what's hot at the big three reflects the explosion of interest in the Internet and the Web.

At Stanford's Center for the Study of Language and Information (CSLI), a partial list of projects of its Interface Laboratory offers the flavor of current work there—the Archimedes Project: Providing Leverage for Individuals with Disabilities, JITLNet: Just in Time Learning on the Internet, Applied Speech Technology Laboratory: To Humanize Speech Technology, and Albots: Autonomous Robots. George White, the director of the Applied Speech Technology Laboratory, is particularly upbeat about the state of the art in speech recognition. With the use of only one hand, White himself is a regular user of speech-recognition systems. (He finds he can achieve 30–50 words per minute with about 90 percent accuracy.) White mentions one project that has him particularly excited: offering powerful speech-recognition servers on the Internet accessible from any telephone—for people with visual impairments or other disabilities or for anyone who can only use a phone. The server would engage the caller in a dialog and could then track down requested information and provide a host of other valuable services.

At Carnegie Mellon's Interactive Systems Lab, the stated goal is "to develop user interfaces that improve human-machine and human-to-human communication." The VERBMOBIL project is a joint venture of the ISL, the German government, and other universities to develop a speech-recognition interface that would essentially be a black box that would sit on a conference table in meetings between Japanese and German businesspeople. The system would capitalize on the fact that both groups already communicate in English. The goal is to provide a sort of language prosthesis that would help both Japanese and German speakers with concepts and expressions they were struggling to convert to English.

The INTERACT Project is described this way: "We seek to derive a better model of where a person is in a room, who he/she might be talking to, and what he/she is saying despite the presence of jamming speakers and sounds in the room (the cocktail party effect). We are also working to interpret the joint meaning of gestures and handwriting in conjunction with speech, so that computer applications ("assistants") can carry out actions more robustly and naturally and in more flexible ways." The project takes a multifarious approach to communications and seeks "to see how automatic gesture, speech and handwriting recognition, face and eye tracking, lip-reading and sound source localization can all help to make human-computer interaction easier and more natural."

MIT's world famous Media Lab, now a venerable ten years old, has always managed to come up with sexy and provocative project titles. Current ones include Smart Clothes, Brain Opera, News in the Future, Things That Think, and Television of Tomorrow. The lab's Autonomous Agents group, headed by Professor Patti Maes, is working on what she calls the "formidable goal" of creating infobots that over time develop a sophisticated understanding of a user's tastes in information and media located on the Internet. In her system, a kind of "digital Darwinism," a number of agents will examine information you've found useful and interesting. They will then disperse into cyberspace to retrieve similar materials. Those who bring back the wrong stuff are culled, and those who find the gold will reproduce, creating agent "offspring" that combine their parents' best features. The idea is that after several generations your agents become more and more intelligent, successful, and useful to you.

Maes's work is still in the early stages, and as she puts it, "Many questions have yet to be answered, others even to be asked."

Patti Maes, director of the Autonomous Agents Group at MIT's Media Lab.

Schematic of Maes's autonomous agents trudging off through an information landscape to find useful data.

CHAPTER SEVEN

The Vector of Desire
The Ultimate Telephone

VR is the first medium that doesn't narrow the human spirit. Remember, we're doing cultural alchemy here. We're introducing a new talisman to Western civilization.

Jaron Lanier, from an interview in Mondo 2000, 1991

irtual Reality (or VR) was the Next Big Thing in 1990. Over the course of a few heady years, VR's chief theoretician, visionary, and inventor, Jaron Lanier, appeared with his trademark dredlocks in virtually every publication in the United States as editors and journalists latched on in waves to the far-out novelty and its implications for media and communications. The VR fever has died down as people realized that imagining how incredible the technology was going to be—though stimulating fun—was not the same as having it widely available next year. A number of applications, chiefly games and simulations, have found their way to the growing VR market.

But right now virtual reality is still a technical novelty that by current standards takes vast amounts of computing power to achieve far less than perfect results. But its powerfully tantalizing implications as a new medium and a springboard to a whole new order of communications persist and drive research interest.

What is it? And why did it seize the imaginations of propeller heads and journalists alike with such force?

Although there are an increasing number of ways to do VR, its trademarks are the large black goggles, glove, and wires everywhere. The goggles are equipped with two tiny videoscreens that offer slightly different perspectives on whatever is "in front of you" to create a 3-D effect. A magnetic device in the goggles or on the top of a helmet transmits information

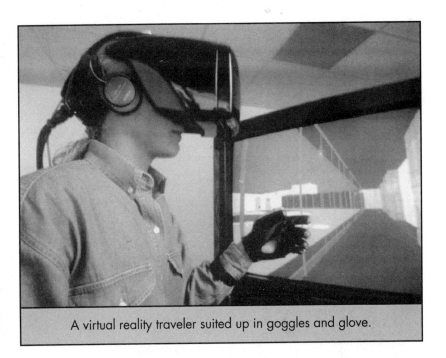

A virtual reality traveler suited up in goggles and glove.

about the "traveler's" head movements to a computer running the VR software. A variety of other sensors may be attached to a "data glove" or even a whole "data suit" so that the traveler looks like someone about to dive into the ocean depths—which is not a bad analogy for the experience.

All information from sensors in the helmet, glove, or suit is translated by the computer into a real-time representation of the traveler in the "virtual world"—a software creation of 3-D graphics representing a house, a palace, the surface of the moon, or just about anything else the designers and programmers can imagine. When you hold out your real hand and point a finger, your virtual hand appears in front of you and points. If you turn to the right in real space, the view alters in virtual space, reorienting you appropriately—and thus you control your journey. The virtual world may also be programmed so that you can reach out and manipulate objects, and walk, fly, or drive through rooms and virtual environments.

In 1968, Ivan Sutherland, dubbed "the godfather of computer graphics," created the first head-mounted display while working at the Uni-

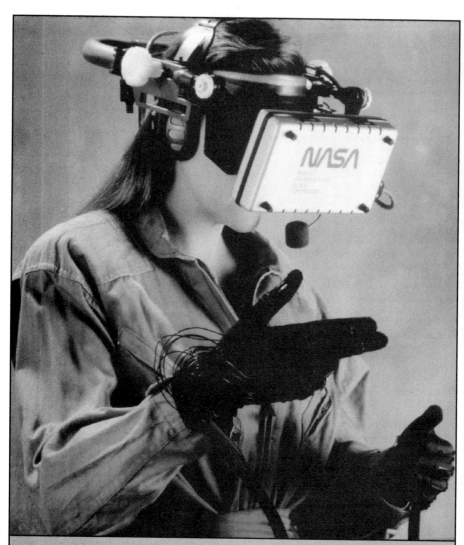

NASA researcher wearing the Virtual Environment Workstation (VIEW). NASA's interest is in developing the system for use by astronauts, allowing them to "explore" the surface of planets without actually being there.

versity of Utah. It was so heavy it had to be suspended from the ceiling. The graphics were crude—"wire frames" representing objects in the virtual world, as with some of the early computer games. In the early 1980s, science fiction writer William Gibson, in a series of short stories and novels, invented a world that involved a global virtual reality network. The books stimulated the imaginations of many young computer scientists, some of whom later became involved in VR development.

NASA became interested in the technology for possible use by remote operators manipulating a robot to build and repair things in outer space. In the late 1980s Lanier and colleagues at his company, VPL Research, Inc., took the technology and ran with it in the directions of whimsy, fantasy, and communications. Unfortunately, Lanier's and VPL's meteoric rise was matched with an even brisker fall.

In a corporate maneuver in 1993 that also stripped him of his VR patents, Lanier was removed from VPL, and the whole field has since taken a back seat to hot interest in the World Wide Web, which became the next Next Big Thing. But somewhere in the second decade of the next century millions may be making the "ultimate" telephone call (see the box "With Bandwidth Enough, and Time . . .).

Although VR became a temporary victim of overblown expectations and overheated imaginations, at a quieter level, it has become by one estimate a $100 million industry. Here are just some of its current uses:

- Flight, tank warfare, and other combat simulations.
- Police training simulating various dangerous situations, such as high-speed pursuits.
- VR shopping: Matsushita Electric has a department store installation that allows a customer to take any of 30,000 appliances from its database and try them on for look and feel in a VR replica of the customer's own kitchen, which the customer can move around in.
- Molecular modeling in pharmaceutical research.
- Public access to virtual entertainment parlors.
- Surgical simulations. With a "virtual cadaver" medical students can perform operations without fear of mangling a real cadaver, or worse, a live patient.
- Home Games: the Mattel PowerGlove, sold for under $100, allows players to do such things as punch bad guys out, virtually speaking.

Future Dark

William Gibson, coiner of the term "cyberspace," is a founder and the best-known practitioner of a subgenre of science fiction sometimes called "cyberpunk."

In his novel *Neuromancer*, Gibson defines his terms:

> Cyberspace. A consensual hallucination experienced daily by billions of legitimate operators, in every nation.... A graphic representation of data abstracted from the banks of every computer in the human system. Unthinkable complexity. Lines of light ranged in the nonspace of the mind, clusters and constellations of data. Like city lights, receding...

In his fiction, Gibson borrows some of the mythos of the cowboy and combines it with the atmospherics and attitudes of the private eye novel and film noir. His hero-not figures are loners and inhabitants of the vast urban stretches of the "Sprawl," a fallen, wildly hi-tech world of the not-too-distant future. If there are any "nice" people, as in AT&T's future vision, we never see them in a Gibson novel. His are dark side characters—quasi- or full-blown criminals who maintain a certain code of honor, or at least high professional standards and skills. They use their computers to perform desperate and literally mind-boggling feats like hijacking heavily guarded industrial secrets and feats of cyber-accounting legerdemain.

Published more than a decade before the advent of the Web, the cyberspace of Gibson's *Neuromancer* is not the chatty, people-filled universe of today's Internet, but rather features the info cowboy loner flying through a vast VR information landscape. Goggles and glove are not needed. The cowboy "jacks into the Net," sending his whole nervous system into cyberspace via a socket surgically implanted in his skull.

Other Gibson "new media," by turns grotesque, intriguing, horrifying, fascinating or all of the above:

> Still from the film *Johnny Mnemonic* showing skull socket and hardware.

- Downloading the "contents" of the brain and personality of an individual before death and granting the individual immortality as a digital being on some recorded medium or at play in the "Net"—free (or condemned) to roam there for eternity.
- Software that you plug directly into your skull socket. The software can give you instant access to special knowledge or physical skills that you want or need. Or it can give you access to the "recorded" experiences and memories of another human being.

Although these kinds of capabilities are now commonplace in science fiction prose and film, the technologies to make them real-world possibilities are not even on the horizon.

30 Years Out—A Roundup

Here, then, is a roundup of ideas about the near future of new media and communications, courtesy of convergences in the entertainment, publishing, and newly deregulated telecommunications industries. These are some of the technologies and their spin-offs that are likely to mature and become ubiquitous over the next 30 years:

- High-bandwidth, high-definition digital television and networked PC's, used for much more than broadcast TV and word processing.
- High-bandwidth communications everywhere in North America and globally in Europe, Asia, and then everywhere else. A fully evolved World Wide Web that reaches into nearly every home, school, and business with interactive multimedia featuring very-high-resolution sound and video.
- The Digital Video Phone: ISDN technology followed by fiber optics makes this long-heralded technology practical and affordable.
- The Virtual Corporation: Also called the "elusive corporation," this is a business entity that may have no more than a few dozen (or less) people at any given location. It conducts its business via the

Net and uses the services of a variety of subcontractors, offering everything from graphic design to accounting services. "Corporate headquarters" is cyberspace.
- World Wide Library On-line: The whole human record, the contents of all the world's libraries and film, video, and music archives available to anyone, anywhere. Increasingly sophisticated and useful search engines make finding what you want, and only what you want, increasingly easy to do.
- Electronic Town Hall and Plebiscite (democracy in cyberspace): It's a dubious notion that abolishing Congress and having all issues decided by electronic plebiscite would be a good idea, even when it becomes possible. While that's being debated, though, political representatives in federal and state governments will utilize the Web to poll constituents and communicate inexpensively with them for fast feedback on issues and pending legislation.
- Cyber U. Lifelong Learning: The University of Phoenix already offers an MBA program on-line. Other universities will follow suit with a variety of courses and degrees. Given the option, most people will want to be educated in the flesh rather than in cyberspace—enjoying the collegiality of high-resolution, real people. But for those who can't conveniently get to a university, the on-line option will be very attractive.
- Networked Schools: The mandate of the Telecommunications Act of 1996 to help schools and other institutions get wired will enable most to have high-bandwidth two-way connections in the first decade of the twenty-first century. An early example of the many kinds of learning that can be done with these connections is oceanographer Bill Ballard's (he explored the Titanic) uplinks and downlinks with students in wired schools across the nation. Ballard's group offers students live visits to current exploration sites. Expect more of the same from astronauts, archaeological and paleontological digs, local, government activities, art openings, and so on. Students and teachers at all levels will have access to the same on-line resources as scholars and scientists.
- Digital Cash: Tangible money—coin, cash, check—is a deeply familiar way to exchange value, but also wasteful of resources, slow, messy, and inconvenient. Expect a gradual cross-over to all digital transactions. With some 300 billion cash transactions per year in the

With Bandwidth Enough, and Time...

Many people who have tried VR have had their "minds blown" with its possibilities for the future. John Perry Barlow was one such enthusiast. In an article for *Mondo 2000* in 1991, he offered these reflections on his experience:

> I have become a traveler in a realm that will be ultimately bounded only by the human imagination, a world without any of the usual limits of geography, growth, carrying capacity, density or ownership. In this magic theater, there's no gravity, no Second Law of Thermodynamics, indeed, no laws at all beyond those imposed by computer processing speed ... and given the accelerating capacity of that constraint, this universe will probably expand faster than the one I'm used to. Welcome to Virtual Reality. We've leapt through the looking glass. Now what? Go ask Alice.

According to those who have spent the most time in VR, including Jaron Lanier, the now-what that makes it truly interesting is the presence (or telepresence) of other people. This suggests that one of VR's key potentials may be as a communications medium, what Barlow has dubbed "the ultimate telephone," another step in removing the barriers between two or more people separated in space.

One of the early Lanier virtual world programs was called Reality Built for Two—wherein two people could meet in a virtual world. A 1990 demonstration sponsored by Pacific Bell featured two suited-up VR travelers playing tag in VR. Their shenanigans in cyberspace were relayed on TV monitors for amazed conferees to watch.

United States alone, there's powerful motivation for some company to come up with a better system (and take a piece of the action, of course). In one such scheme, you'll carry a new type of card with an embedded chip that can store 100 times more information than the current magnetic strip cards and also can offer encryption. You may

The demonstration offered a hypothetical look at a whole new telecommunications future: What if you could punctuate or enhance your telephone conversation with a true, or virtually true, meeting. Take a walk around the Taj Majal while talking. For business calls, a meeting in one of your company's well-appointed "virtual conference rooms." For family calls, some favorite hangout: a summer place, a park, a favorite game. For lovers, a favorite beach or bistro. For gamers, adventures of every description.

Musicians could meet to jam on replicas of real or imaginary instruments. Actors could improvise or rehearse a scene. Add your VR fantasy here.

Another potential technology that could dramatically enhance the illusion of "reality" in VR would be the addition of feedback and resistance devices in the data glove or suit. When you touch someone or something in a virtual world, you would then "feel" it. VR experimenters have demostrated an early stage of this technology. Howard Rheingold, author of the book *Virtual Reality*, has even volunteered a name for feedback VR—"Teledildonics"—a term sure to intrigue sophisticates and further alarm the U.S. Congress.

Before all these wonders (or horrors) come to pass, though, a few little problems will need to be solved. Because a complex VR world taxes even a powerful computer to its limits, there can be a slight lag between what you do in VR and the program's response. In part due to this, many people in VR experience a kind of motion sickness and disorientation. Add to this the bulk and discomfort of the current equipment and we may be 20 years or so away from an application of VR as widespread as the telephone.

be able to "fill up" such a card at a banking machine, or even on your home computer, and you'll use it in lieu of cash anywhere by simply waving it at a reading machine. It may be read while still in your wallet. Naturally, any such system will have to minimize invasions of personal privacy and maximize security.

"Says Saffo"

Another professional future watcher who pointedly resists the job description "futurist" is Paul Saffo, a director of the Institute for the Future, in Menlo Park, California. Preferring the term "forecasting" to "prediction," Saffo describes himself as a "specialist in the long-term commercial impact of new information technologies who devotes much of his time to making sense of structural shifts occurring at the intersections of computer, consumer electronics and communications industries." Although in many ways he keeps a lower profile than some of the other information gurus, "says Saffo" may be the most commonly found word pair in the nation's business reportage on new technologies.

Paul Saffo: "Never mistake a clear view for a short distance."

In spite of, or because of, his company's professional interest in the future, Saffo has made the study of the past an important part of his work. In one of his talks, Saffo is likely to spend as much time on the history of technology as on the future, drawing on past follies and wonders to elucidate current trends.

Like some other

prognosticators, he is fond of the aphorism; some of his favorites include:

"Never mistake a clear view for a short distance."
"Change occurs at the speed of thought."
"Everything takes 20 years to become an overnight success."
"Technology does not drive change at all, technology only enables change."
"It takes 30 years to turn a technology into a medium."

Saffo reminds us that most future visions never come to pass as expected. Bell's first vision for the telephone was a means to pipe musical entertainment to towns too small to have their own live theaters. In Saffo's view, the first thing we do with a new technology, after oohing and ahhing over it, is to "pave the cowpaths"—use it in fairly unoriginal ways to do something that's already being done. The phonograph was going to be a spoken newspaper. Radio was to be a wireless telephone like the ham system, until someone noticed that lots of people enjoyed eavesdropping on transmissions. Although it had been demonstrated successfully before the turn of the century, the first commercial radio broadcast didn't occur until 1922.

Saffo stresses that just because we can see what's coming doesn't mean it will be here any time soon—or that it will all arrive at once. Among other things, this perspective has a cooling effect. It's good medicine for the hyperventilation and tachycardia that can come from becoming too entranced by the future.

Although we're fond of speaking of the "speed of change," Saffo ascribes the sensation to quantity, not velocity. "What's giving us the acceleration effect," he says, "is not that things are changing more quickly, but that more things are changing at the same time. It is the cross impact of those many changes that is creating all the turbulence we feel in our lives."

His antidote, quoting the motto of the Aldine Press of 1501: "Make haste slowly."

- Smart Home, etc: Bill Gates is pioneering this with his immense home on Lake Washington in Seattle. Lights turn on and off as people leave and enter rooms. The house knows who and where you are by a little clip-on device and routes your incoming calls, favorite music, and works of art (on immense flat-technology wall screens) to the appropriate room in the house. Your car, which is in communication with your house for alarms and other signals, is equipped with GPS and map software to show you where you are, where the traffic jams are, where you're going, and the best way to get there at all times. The car can give the precise (and always patient) answer to children whining, "When are we going to get there?"
- Smart transportable communications devices like the personal digital assistant with video. We'll carry these devices in pockets, purses, and briefcases or wear them as ornaments, on our wrists, around our necks, in our ears, and in our glasses.
- Information Warfare: A recent article in *AirPower* magazine outlines the principals of a new emphasis in warfare:

  Information warfare relies on sophisticated communication, imbedded intelligence, access to space, and real-time decision loops. It is permeated by information feeding precision weaponry, multispectral sensors providing real-time data about the battlefield, and tightly woven command and control of combined arms elements.

  More important than the brawn of new weapons systems will be brains of communications and information systems—enhancing one's own and eliminating an enemy's ability to communicate and gather information, described as "electronic decapitation" and "sensor denial."
- Cyber crime and terrorism: For the same reasons that military strategy is changing to reflect the importance of real-time information and communications, the Internet and telecommunications systems will interest terrorists of all stripes. Though the Internet was designed to withstand attack, it is by no means invulnerable to damage and mischief. Crime and crime-stopping will be as much a part of cyberspace as they are in contemporary "off-line" society.
- Virtual Reality communications: Only as processing power increases enough to provide smooth sailing in cyberspace will we

know whether VR is going to be "the ultimate telephone," or find other valuable and entertaining uses that become widespread.
- Natural Language Understanding (NLU) and Continuous Speech Recognition: True NLU will continue as a holy grail of artificial intelligence research. There will be no breakthrough that suddenly delivers up a computer that's a congenial conversationalist, but as processing speeds and power advance via Moore's law, computer scientists will find cunning ways to make computers steadily more responsive to typed-in language and speech.
- Intelligent, Intelligent Agents: To an important degree, the advent of truly "intelligent" agents is linked to the fate of natural language understanding. In the meantime, anything that helps us use the information infrastructure intelligently and easily is an "intelligent agent." Navigational aids and search engines will be in a constant state of evolution as computing power increases and becomes cheaper and ingenious new strategies are developed. Agents could be software you buy or download or part of a powerful file server you access to help you accomplish what you want on the Net.
- ??? The Five-Year Surprise: With so many technologies and industries advancing at once, unpredictable synergies in new media and communications are likely to produce something that catches nearly everyone by surprise every five to ten years. From now on.

The prospect of these new media individually and in the aggregate is astonishing, exciting, alarming, baffling, and more, depending on one's mood and point of view. None of these technologies offers us a prescription for utopia, or even an automatically better world, although they will result in enormous new powers for individuals, communities, and states. The instant access to new media resources and improved telecommunications with our fellow humans is attractive. The stress and dislocations of change and the potentials for abuse and loss of privacy are not so attractive. Technological optimists hope and believe that if technology gets us into trouble, other technology can help get us out. The pessimists are not convinced. Like it or not, though, for better and for worse, interesting times will continue.

TIMELINE

1450   Johann Gutenberg invents movable type and the printing press and becomes the founder of printing in Europe.

1456   Johann Gutenberg's first book, a Latin Bible, is published. The inventor dies in poverty and obscurity, but his invention of the printing press has an immense and incalculably diffusive affect on all subsequent history.

1490   Aldus Manutius founds his press in Venice. It becomes one of the first publishing houses, developing the first "modern" book with legible italic typefaces, page numbers and other improvements, and the "pocket"-sized book. The house publishes over a thousand editions over the next century.

1609   One of the first regularly published weekly newspapers appears in Strasbourg.

1714   The first writing machine with raised letters is patented.

1791   The First Amendment to the U.S. Constitution guarantees freedom of the press.

1794   The optical telegraph system, invented by Claude Chappe, is used to connect Paris and Lille.

1822   The Frenchman Nicephore Niepce makes the first photograph. Seven years later he becomes an associate of Jacques Daguerre of daguerrotype fame, the first commercial use of photography.

1837   The first operating web-fed rotary press comes into use.

1840   Samuel Morse patents his electric telegraph.

1843   The first successful electromechanical fax machine is patented by the Scottish inventor Alexander Bain.

1844   The Chappe optical telegraph connects 29 cities through 500 stations.

1865   James Clerk Maxwell publishes his electromagnetic theory of light.

T I M E L I N E

1874	The first practical typewriter, Christopher Sholes's "Remington No. 1," is brought to market. The first typed memo is produced shortly thereafter.
1876	Alexander Graham Bell invents the telephone—analog transmission. Voltage impressed on the line was proportional to the sound pressure at the microphone, beginning a century of analog transmissions.
1876	The telegraph system is huge, the largest company on the planet, with 214,000 miles of wire and 8,500 telegraph offices. There are specialized telegraphs for businesses, stock traders, government, police, and fire departments. The head of Western Union calls Bell's invention "an electrical toy."
1878	Thomas A. Edison develops a cylindrical phonograph.
1880s	Photoengraving and halftone screen—a range of blacks and grays—come into use.
1884	Paul Nipkow invents the electromechanical scanning disk, a device used in early television research.
1887	The linotype machine is invented by Ottmar Mergenthaler, a keyboard-operated machine that composed and cast a justified line of type—used primarily for newspapers. The monotype keyboard produced a punched tape that instructed a separate typecaster to produce and set type.
1890	Having started in Boston among the technically curious and well-to-do, the telephone network soon spreads throughout New England. By 1904, it's all over the continent.
1890s	Halftone color printing becomes possible.
1892	The first automatic phone network switch demonstrated.
1893	Edison invents the kinetiscope, a primitive form of motion picture projector.
1895	Guglielmo Marconi transmits wireless signals for 1 mile.

T I M E L I N E

1895	Sir William Crookes invents the cathode-ray tube, which becomes the basis for television.
1897	J. J. Thomson proposes his electron theory.
1900	The American scientist R. A. Fessenden broadcasts speech for over a mile with a wireless.
1904	Sir John Fleming invents an electronic vacuum tube, to be used in detecting, modifying, and amplifying electromagnetic waves.
1907	Lee De Forest patents his "audion"—a three-element vacuum tube that was used in radar, television, radio, and other electronic devices until it was replaced by the transistor decades later.
1910–20	D. W. Griffith develops the "grammar" of film (close-up, medium, and long shots; flashbacks and jumps in time) to tell cinematic tales in ways very different from the stage play.
1913	There are 92,000 telephones in use in Paris; 500,000 in New York.
1913	AT&T under Theodore Vail greatly improves its long distance lines. Vail's corporate slogan, "One Policy, One System, Universal Service" becomes the AT&T long-term goal. The network now consists of open wire systems transmitting a dozen simultaneous conversations through the use of frequency-division multiplexing of the 4 kz analog voice band channels. Switching is done manually by phone operators whose ranks are growing exponentially.
1920s	After 50 years of the telephone, it becomes possible to place a call between any two phones in the country (though still a luxury).
1923	Vladimir Zworykin invents the "iconoscope" and the "kinescope" for transmitting and receiving television images.
1927	Philo Farnsworth patents his version of an all-electronic TV system.
1938	Electrostatic copying is invented by Chester Carlson.

TIMELINE

1941	Five years after London and Paris, the first transmissions of regularly scheduled TV programming begin in the United States, with broadcasts from the Empire State Building.
1945	Vannevar Bush, in an essay "As We May Think," imagines and describes the "memex," a multimedia information storage and retrieval system using associational connection as well as indexing. It is extrapolated from wartime and cutting edge business technologies.
1945	Arthur C. Clarke describes his ideas for a communications satellite in geosynchronous orbit in an article in *Wireless World*.
1948	Claude Shannon's paper "A Mathematical Theory of Communication" creates a philosophical revolution in the way information and communications were understood.
1948	Bardeen, Brattain, and Shockley invent the transistor, replacing the vacuum tube, which evolved to become the basic component for modern electronics.
1950s	Toward the end of the decade telephone traffic begins to be carried through beams of radio transmission, tower to tower across the country, carrying thousands of conversations.
1959	Plain paper copying is developed by the Haloid Corporation (later to become Xerox).
1959	Robert Noyce, co-founder of Intel, combines several electronic components on a silicon chip to create the first primitive integrated circuit. Thirty-five years later Intel squeezes 6 million components on a single chip.
1961	Bell System installs the first digital carrier T-1 system.
1962	Twenty years after the first regularly scheduled TV programming had begun, North America has 59 million TV sets and Europe 26 million; and 65 countries are transmitting regular programming.
1962	The United States launches the Telstar satellite, the first commercial satellite to provide active relay of live TV and voice

TIMELINE

communications between the United States and Europe. Television frequencies (line of sight) now extend across continents.

1963 The space probe "Mariner II" maintains radio contact with earth from 36 million miles out.

1964 The modern fax machine arrives through long-distance xerography.

1964 *Understanding Media: The Extensions of Man,* by Marshall McLuhan, is published.

1964 Douglas Engelbart invents the Mouse, windows, and linked documents.

1967 Theodor Nelson describes Hypertext—notes, documents, and other media digitally linked to each other. A quarter-century later hypertext becomes the primary user interface of the World Wide Web.

1969 ARPANET, the predecessor of the Internet, is started with funds from the Defense Department's Advanced Research Projects Agency.

1970s AT&T finally achieves its goal of universal telephone service.

1970s IBM develops the Magnetic Tape Selectric Typewriter, precursor to the word processor.

1970s Alan Kay coins the term "personal computer" and describes the Dynabook, a small, book-sized computer that has many features of today's notebooks.

1970s The Alto computer is developed at Xerox Palo Alto Research Center. The brainchild of Alan Kay and other scientists, the computer uses a mouse and graphical user interface, both adopted and extended in the Macintosh computer and Windows software.

1970s Theodor Nelson and colleagues develop the concept of Xanadu, a "universal electronic publishing system and archive."

TIMELINE

1971	The first optical fibers with a transparency suitable for communications use are made.
1973	Vinton Cerf and Robert Kahn design the foundations of what will become the Internet.
1976	Electric Pencil, the first modern word processing program, is written by Michael Shrayer for the Altair, a microcomputer for computer hobbyists.
1976	Stephen Wozniak and Stephen Jobs design a prototype that becomes the Apple I Computer, the first ready-to-use personal computer for the public at large.
1979	The laser xerographic printer is developed by IBM and Xerox.
1981	Bell System installs the light-wave, fiber optic "Northeast Corridor" from Boston to New York to Philadelphia and Washington.
1984	The Bell System is broken up by court order into the Regional Bell Operating Companies, or "Baby Bells."
1989	AT&T's "long-haul network" is nearly all digital.
1990–91	Tim Berners-Lee and colleagues develop the software and protocols for the World Wide Web at CERN, the European particle research center.
1993	The World Wide Web has 143 file servers. Two years later there are 40,000.
1996	The Telecommunications Reform Act of 1996 opens the new media floodgates.

FURTHER READING

Carlson R. and Goldman B. *Fast Forward: Where Technology, Demographics, and History Will Take America and the World in the Next 30 Years* (New York: Harper Business, 1994).

Crevier D. *AI: The Tumultuous History of the Search for Artificial Intelligence* (New York: Basic Books, 1993).

Dreyfus H. and Dreyfus S. *Mind Over Machine: The Power of Human Intuition and Expertise in the Era of the Computer* (New York: Free Press, 1986).

Dreyfus H. *What Computers Can't Do: The Limits of Artificial Intelligence* (New York: Harper & Row, 1972).

Flatow I. *They All Laughed . . . From Lightbulbs to Lasers* (New York: HarperCollins, 1992).

Freedman D. *Brainmakers: How Scientists Are Moving Beyond Computers to Create a Rival to the Human Brain* (New York: Touchstone, 1994).

Freiberger P. and Swaine M. *Fire in the Valley* (Berkeley, Calif.: Osborne/McGraw-Hill, 1984).

Kelley K. *Out of Control: The New Biology of Machines, Social Systems, and the Electronic World* (New York: Addison-Wesley, 1994).

Kurzweill R. *The Age of Intelligent Machines* (Cambridge Mass., MIT Press, 1990).

Lambert S. and Ropiequet S, eds. *CD-ROM: The New Papyrus* (Microsoft Press, 1986).

Leebaert D. *Technology 2001: The Future of Computing and Communications* (Cambridge, Mass.: MIT Press, 1991).

Levy S. *Insanely Great* (New York: Viking, 1994).

McLuhan M. *Understanding Media: The Extensions of Man* (Cambridge, Mass.: MIT Press, 1995).

McLuhan M. & Powers B. *The Global Village: Transformations in World Life and Media in the 21st Century* (Oxford: Oxford University Press, 1989).

Negroponte N. *Being Digital* (New York: Knopf, 1995).

Pinker S. *The Language Instinct: How the Mind Creates Language* (New York: Harper Perrennial, 1995).

Rheingold H. *The Virtual Community: Homesteading on the Electronic Frontier* (New York: Addison-Wesley, 1993).

Rheingold H. *Virtual Reality: The Revolutionary Technology of Computer-Generated Worlds—and How It Promises to Transform Society* (New York: Touchstone, 1991).

Roszak T. *The Cult of Information* (Berkeley: University of California Press, 1994).

Scientific American. 150th Anniversary Issue, September 1995.

Shurkin S. *Engines of the Mind: A History of the Computer* (New York: W.W. Norton, 1984).

Sterling B. *The Hacker Connection* (New York: Bantam Books/Doubleday, 1992).

Stoll C. *The Cuckoo's Egg* (New York: Pocket Books/Simon & Schuster, 1990).

Turkle S. *The Second Self: Computers and the Human Spirit* (New York: Simon & Schuster, 1984).

Vaughan T. *Multimedia: Making It Work* (Berkeley, Calif.: Osborne/McGraw Hill, 1994).

WIRED Magazine.

INDEX

ADSI (Asynchronous Digital Subscriber Line), 61
Alto computer, 47–48
America Online, 96, 98
Apple Computer, 106, 111
Archaeotype, 83–84
ARPAnet, 65–66
artificial intelligence (AI), 102–119
AT&T, 35, 52, 53–54, 56, 58, 94, 106–108, 111
ATM (asynchronous transfer mode), 61–62
automatic speech recognition (ASR), 109–111
avatars, 69, 70

Baird, John Logie, 37–38, 39
Ballard, Bill, 127
bandwidth, 59–60
Bardeen, John, 37
Barlow, John Perry, 9, 99, 100, 101
Bartlett's Familiar Quotations, 84
Bell, Alexander Graham, 31, 32, 33, 54, 104, 131
Bell System, 33, 52
Birkerts, Sven, 75, 99–100, 101
bit, 59
bit-mapping, 48
bozo filter, 96
Brand, Stewart, 91
Brattain, Walter, 37
Brill, Stephen, 84
bulletin board systems (BBSs), 66
Bush, Vannevar, 35–37

cable television, 62, 64
Carnegie Mellon University, 45, 117–118
cash transactions, digital, 127–129
CD-ROM, 81–83, 84–85, 86, 87, 89
cellular phone, 57–58
Chappe, Claude, 29
Chou, Luyen, 83–87
Clarke, Arthur C., 56
communications, evolution of, 27–42
Communications Decency Act, 98
computer, development of, 42–48
computer-mediated communication (CMC), 103
continuous speech recognition (CSR), 105, 111–112, 133
copyright protection, 90
crime, 95–96, 132
Cyberspace, defined, 124
CYC, 115–116

Dalton School, 83
data compression, 60
Deep Blue, 104
dictation software, 108, 111

digital editing, 80
digital revolution, 8–9
 convergences in, 11
 forces driving, 9–11
 future of, 126–133
 personal effects of, 99–101
 public reaction to, 12
Dreyfus, Hubert, 103, 104
DVD (digital video/versatile disk), 82
Dyson, Esther, 92–93

Edelcrantz, Abraham, 27
Edison, Thomas, 33
education, 25–27, 83–87, 127
e-mail, 20, 59, 67
Engelbart, Douglas, 43, 44–45, 47, 89
ENIAC computer, 13, 37
expert systems, 105

Farnsworth, Philo T., 38
fiber optics, 10, 54–56
file transfer protocol (FTP), 66–67
film, 76–77
film editing, 79–80
fire, taming of, 49–50
flaming, 96

Gates, Bill, 42, 48, 70, 71–72, 75, 132
Gibson, William, 123, 124
Gilder, George, 67, 70–71, 72–73, 91–93
Gleick, James, 96
Gray, Elisha, 31
Green's Axiom, 12

hackers, 33, 95–96
Harper's Magazine, 99
Hawking, Stephen, 109
home, smart, 132
hypermedia, 36
hypertext, 89

idiot savants, 115
information superhighway, 10, 71, 73–74
information warfare, 132
intelligent agent, 19–20, 116, 133
interactive digital multimedia, 78
Internet, 10–11, 64, 132
 business over, 89, 91–93
 computer games, 86–87
 crime on, 95–96
 ethics and, 96–97
 future of, 72–73
 origins of, 65–66
 pornography on, 97–98
 size of, 68

INDEX

Internet *(continued)*
  speech recognition servers on, 117
  utilities of, 66–67, 69–70
Internet relay chat (IRC), 67, 69
ISDN (Integrated Services Digital Network), 60–61, 126

Jenkins, Charles Francis, 39
Jobs, Steve, 42, 48

Kaufman, John, 97
Kay, Alan, 42, 46–47
Kelly, Kevin, 99, 100
Kurzweil, Raymond, 105, 110

Lanier, Jaron, 102, 120, 123, 128
learning technologies, 83–86
Learning Technologies Interactive (LTI), 84
Lenat, Doug, 115, 116
Levy, Steven, 43
Lucky, Robert, 52

machine vision, 105
Maes, Patti, 118–119
Markoff, John, 95
Massachusetts Institute of Technology (MIT), 45, 117, 118–119
McCarthy, John, 102
McLuhan, Marshall, 8, 12, 16, 40–42, 75, 78, 98
Memex, 37
mergers, 11, 72
Metcalfe's law, 72
Microsoft, 43
Minitel system, 64
Minow, Newton, 40
Minsky, Marvin, 103
Mitnick, Kevin, 95–96
Moore, Gordon, 9
Moore's law, 9, 10, 14, 72, 111
Morse, Samuel, 30
Morse Code, 30
mouse, 44–45, 47
MUD object orienteds (MOOs), 69, 70
multimedia composition, 23–24
multi-user adventures (MUAs), 69
multi-user dungeons (MUDs), 69–70
musical composition, 78–79, 103, 104
musical signatures, 21

nanobucks, 73, 91
national information infrastructure (NII), 71
natural language understanding (NLU), 105, 112–114, 133
NAVSTAR Global Positioning System, 57
Negroponte, Nicholas, 62, 63, 72–74

Nelson, Ted, 73, 89, 91
Net-Tops, 73
Neuromancer (Gibson), 124
Newell, Alan, 102–103
New Laboratory for Teaching and Learning, 83–84
Nipkow, Paul, 38

Oikarinen, Jarkko, 67
on-line communities, 10–11

Palo Alto Research Center (PARC), 46–48
Peppel, Tyler, 82
personal computer (PC), 14–15, 37, 93, 126
  Web, 73, 94–95
personal digital assistants, 58, 132
pornography, on Internet, 97–98
publishing, 90–91, 92–93

Quin: Tomb of the Middle Kingdom, 84–85

Rheingold, Howard, 66, 69
robotics, 105

Saffo, Paul, 64, 130–131
satellite system, 56–57
Shimomura, Tsutomu, 96
Shockley, William, 37
SHRDLU, 113–114
signatures, multimedia, 21
Simon, Herbert, 102–103
Slouka, Mark, 100–101
software, 9, 92–93
speech recognition, 36, 117
  automatic (ASR), 109–111
  continuous (CSR), 105, 111–112, 133
speech synthesis, 105, 108–109
Stanford University, 45, 117
Sterling, Bruce, 33, 53
Stoll, Clifford, 95
student-teacher explorer teams, 26
Sutherland, Ivan, 121, 123
Swinton, A. Campbell, 39
switchboard, 51–52
switching systems, 52–54

Talbott, Stephen, 98
talk invitation, 67
Telecommunications Reform Act of 1996, 11, 62, 70, 97, 98, 127
telegraph, 29–31
telephone, 75, 94, 126
  cellular, 57–58
  fiber optics, 54–56
  invention of, 31–35

INDEX

ISDN (Integrated Services Digital
   Network), 60–61
  Minitel system, 64
  switching systems, 50–54
  wireless systems, 58–59
  wristphone, 58
television, 126
  cable, 62, 64
  development of, 37–39
  video on demand, 62–64
Thurber, James, 8, 16–17
Toffler, Alvin and Heidi, 102, 103
Tomlin, Lily, 51
track ball, 44
transistor, 37
Turkle, Sherry, 65
twisted pair, 55

Universal Resource Locator (URL), 89
UNIX, 65, 66
Usenet, 66

Vail, Theodore, 35
video
  CD-ROM, 81–83, 84–85, 87, 89

conferencing, 21–23
editing, 79–80
learning technology, 83–86
networked, 86–87
video on demand, 62–64
virtual corporation, 126–127
virtual reality (VR), 120–126, 128–129, 132-133

Watson, Thomas, 31
Web PCs, 73, 94–95
webtops, 25
White, George, 117
Whitman, Walt, 16, 816
Windows, 43–44
Winograd, Terry, 112–113, 114, 115
World Wide Web, 10–11, 25, 60, 64, 67, 68, 70,
   73, 74, 86, 87–89
Wozniak, Steve, 42

Xanadu, 90–91
xerography, 35
Xerox, 46, 48

Zworykin, Vladimir, 38, 39

PHOTO CREDITS

Photo Research: Beth Krumholz, Christopher Deegan, Laurie Platt Winfrey, Carousel Research

p.2—© Michael Simpson/FPG International
p.3—Wide World Photos
p.4—©Tony Stone Images/David Young Wolff

INTRODUCTION
p.10—Source: Reprinted by permission of FORBES Magazine, ©Forbes Inc. 1995
p.11—TOLES © The Buffalo News. Reprinted with permission of UNIVERSAL PRESS SYNDICATE. All rights reserved   p.13—Courtesy of IBM Archives

CHAPTER 1
p. 20—©Tony Stone Images/Mark Lewis   p.22—©Jeff Titcomb/Liaison International
p.25—©Tony Stone Images/David Young Wolff

CHAPTER 2
p. 28—Woodcut from Jean de Mandeville's Travels in the Orient, 1481.   p.30 left—Courtesy Smithsonian.   p.30 right—©AP/Wide World Photos   p.32—©AP/Wide World Photos   p.34—©Corbis-Bettmann   p.36—Courtesy of the MIT Museum   p. 38—©UPI/Corbis-Bettmann   p.39—©Corbis-Bettmann   p.41—©UPI/Corbis-Bettmann   p.45 left—Courtesy of Bootstrap Institute, Freemont, CA   p.45 right—Courtesy of SRI International, Menlo Park, CA
p.46—Courtesy of Xerox Corporation

CHAPTER 3
p.51—©Photofest   p.53—©Hank Morgan/Science Source/Photo Researchers   p.54 left and right—©Tony Freeman/PhotoEdit   p.57—©Mark Richards/PhotoEdit   p.58—Courtesy of Scientific American and AT&T   p.64—©Alan Oddie/PhotoEdit

CHAPTER 4
p. 68—Courtesy A.M. Rutkowski, General Magic, Inc.   p.69—Courtesy of CompuServe, Inc.
p.71—©AP/Wide World Photos   p. 72—Courtesy Gilder Technology Group, Inc.
p.74—©Bourg/Liaison USA

CHAPTER 5
p.79 above—©Alexis Duclos/Gamma Liaison   p.79 below—©Raphael Gaillarde/Gamma Liaison
p.80—©Spencer Grant/Liaison International   p.84—Courtesy of Learn Technologies Interactive
p.85 above and below—Courtesy of Time Warner Inc.   p.88 above—©1996 Arts International, a division of the Institute of International Education. Design: Rebecca Lown, OVEN Digital, NYC
p.88 below—Courtesy of OVEN Digital, NYC   p.93—Courtesy of EDventure Holdings
p.99—Courtesy of Electronic Frontier Foundation   p.100—Courtesy of WIRED magazine

CHAPTER 6
p.111—Courtesy of IBM   p.114—Courtesy of Terry Winograd, Stanford University
p.119 above—Courtesy of MIT Media Lab   p. 119 below—Courtesy of MIT Media Lab and Scientific American

CHAPTER 7
p.121—©P. Howell/Liaison USA   p.122—©Photo Researchers   p.125—©Photofest
p.130—Courtesy of the Institute for the Future